CW01160075

CROWOOD COLLECTORS' SERIES

Bang & Olufsen

TIM JARMAN & NICK JARMAN

THE CROWOOD PRESS

First published in 2008 by
The Crowood Press Ltd
Ramsbury, Marlborough
Wiltshire SN8 2HR

enquiries@crowood.com

www.crowood.com

This impression 2020

© Tim Jarman and Nick Jarman 2008

All rights reserved. No part of this publication may be reproduced or transmitted in any form or by any means, electronic or mechanical, including photocopy, recording, or any information storage and retrieval system, without permission in writing from the publishers.

British Library Cataloguing-in-Publication Data
A catalogue record for this book is available from the British Library.

ISBN 978 1 84797 068 8

Typeset by Jean Cussons Typesetting, Diss, Norfolk
Printed and bound in India by Replika Press Pvt. Ltd.

CONTENTS

ACKNOWLEDGEMENTS		4
FOREWORD		5
INTRODUCTION		6
1	1925 TO 1964: THE EARLY YEARS	9
2	1964 TO 1969: NEW MARKETS, NEW TECHNOLOGIES	22
3	1970 TO 1979: A DECADE OF ACHIEVEMENT	49
4	1980 TO 1989: THE MICROPROCESSOR AGE	90
5	1990 TO 2000: THE SURVIVAL YEARS	145
6	PRACTICAL COLLECTING	179
7	SYSTEM BUILDER	209
INDEX		222

ACKNOWLEDGEMENTS

Thank you to everyone who allowed us to photograph their B&O equipment for this book. Many items shown in this book are from the authors' own collections, but Neil Briffett, Alan Jewell, Mike Kemp, Mike King and Stewart Clarke all supplied items without which this book would not have been complete. Thank you also to Serena Dolby for proofreading and providing many valuable comments, and to our parents David and Monika for providing a location at which to take many of the photographs.

Finally thank you to our wives, Revai and Marion, for their support and encouragement throughout the project.

Tim Jarman and Nick Jarman
www.beocentral.com

FOREWORD

Bang & Olufsen occupy a unique position in the audio and home entertainment community. I can think of few other manufacturers who remain popular, sometimes obsessively so, among music and technology enthusiasts but who, at the same time, are a familiar name to the general public – often being a name that they know for style and pedigree; and a name to which many greatly aspire.

What is less well known by some, however, is just what a long and illustrious history the company has, stretching back to 1925, and the innovations that they have introduced, and even helped to develop, along the way. As a result there is a bewildering array of equipment out there bearing the Bang & Olufsen name and a helpful guide to this is most welcome.

Author Tim Jarman is a well known name in vintage Bang & Olufsen circles; it is often said about the company or their products that, if he doesn't know it, it isn't worth knowing. Having experienced his knowledge and abilities first-hand on my own Beogram CD 50 CD player, I have no reason to disagree with this! He and his brother Nick are already known for their internet fount of B&O knowledge, www.beocentral.com, alongside Tim's vast collection of B&O's most important models. A historical account of the most noteworthy models was long overdue, and I can think of no-one better equipped to write it.

Consequently, if you are new to the marque, wondering which product to spend your hard-earned money on, an enthusiast looking to expand your knowledge, or an owner wondering which items would match best with the equipment you already own, then you need look no further. Be warned, though, Bang & Olufsen equipment is very addictive, and you could well find yourself with a considerable collection of your own before too long...

Adam Smith, *Hi-Fi World Magazine*

INTRODUCTION

Bang & Olufsen is one of the great survivors of the radio and television industry. Almost as old as entertainment broadcasting itself, yet in recent years seen as unassailably modern, as a small company from a small country, their endurance and the depth and breadth of their product ranges are both remarkable and unique. Along with Lego bricks, bacon and fine butter they are one of Denmark's most famous exports and as such are also great ambassadors for European design the world over.

Perhaps the greatest number of people who know Bang & Olufsen will associate them most strongly with the expensive, sleek home entertainment products that they stunned the buying public with year after year throughout the 1970s and 1980s. An interest may have roots in the marvellous equipment owned by a wealthy (or perhaps just discerning) parent or relative, or the experience may have been firsthand with an exciting purchase that led to many years of pleasure.

In recent times, all types of electronic equipment have become disposable, designed along the whims of fashion and built to last just a year or two. While this approach no doubt suits the long-term aims of certain sections of the industry, it has lead to a generation of products that do not satisfy in the same way as the models of past decades. Perhaps it is as a result of this that an interest has grown in collecting and restoring old television, radio and audio equipment. Collecting Bang & Olufsen equipment has taken time to become

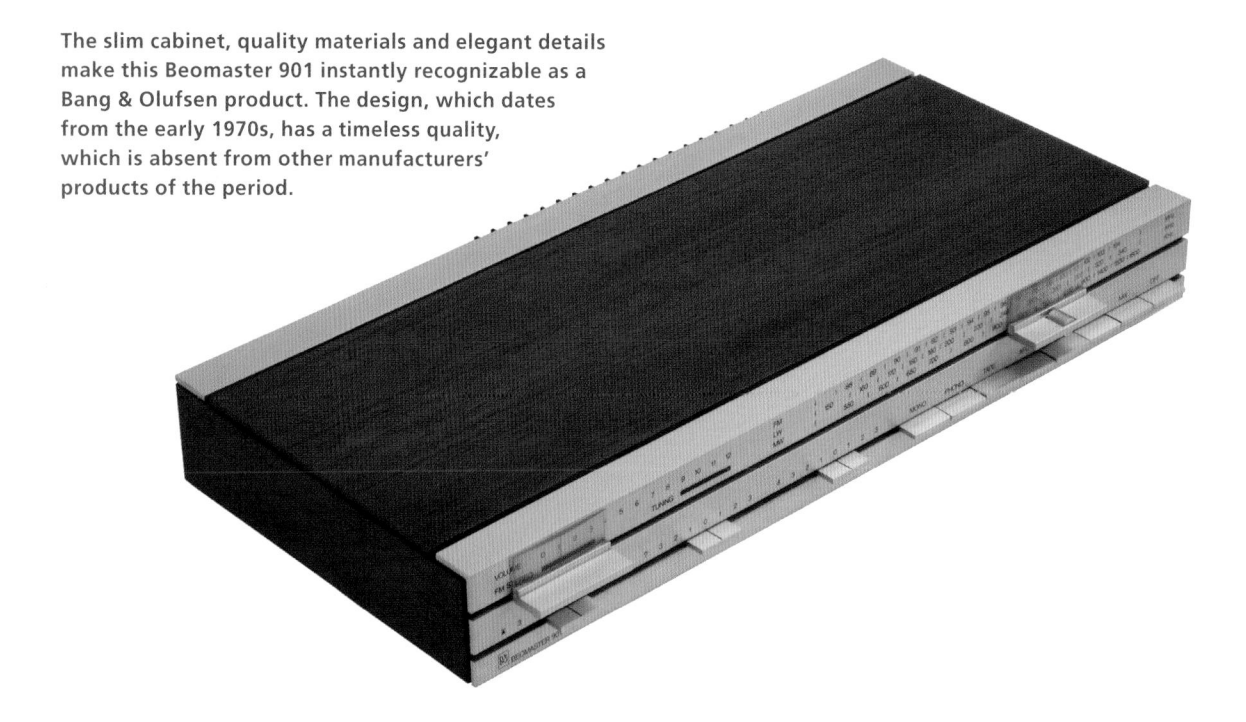

The slim cabinet, quality materials and elegant details make this Beomaster 901 instantly recognizable as a Bang & Olufsen product. The design, which dates from the early 1970s, has a timeless quality, which is absent from other manufacturers' products of the period.

a real aspect of this movement, this is probably due in no small part to the fact that that the models from the period of interest have had such long lives as working items with their original owners and it is only recently that they have been available in large quantities on the collector's market. Another problem is that audio equipment in particular works best when it is assembled into systems of matching items. However, these can be difficult to find, operate and understand. The Bang & Olufsen range has always been large, and knowing what goes together is a complex subject about which there is little information available. One of the aims of this book is to begin to fill this gap.

The first five chapters tell the story of Bang & Olufsen from its foundation as a company in 1925 up until the end of the year 2000. The story is told through products rather than personalities because products represent the most accessible part of the company and are the part that the collector will have to deal with first hand. The reason for selecting 2000 as the particular year for a cut-off date is that at the time of writing, products made any more recently are really rather too new to be considered collectable and, in the main, can be bought through Bang & Olufsen dealers as part of their 'Second Life' reconditioning scheme. Throughout these chapters, the reader will find detailed photographs of selected models, all taken especially for this book. The final two chapters deal with the practical aspects of collecting, and contain help and advice for both the beginner and the more experienced, including tables that show which models work well together.

This book is not intended as a technical repair manual. Electronic repairs of any type are complicated to do and require finely tuned theoretical and practical skills. These can be learned, but the subject is difficult to master and one book alone cannot hope to give even a workable grounding to the complete novice. It is unfortunate that, as the mainstream consumer electronics market becomes ever more centred on disposal rather than repair, the high street shops to which defective television and audio equipment used to be able to be entrusted are now a vanishing sight, just as the interest in older models really starts to gain popularity, but one can only live in the world as it is.

The only advice that can be offered to those who wish to bring broken items back to life is either to learn the skills or to find someone who has them and treat them kindly.

As a collector, it pays to see the acquisition of equipment differently to someone who is either buying something new or second-hand just to use it for its intended purpose. The collector is free to pursue their interest without regard to the various whims that pass through the electronic marketplace; in fact they can take advantage of them. At the time of writing, it is fashionable to replace traditional television sets with flat-panel types and, as a result, many fine older models (and in some cases not that old at all), sometimes in excellent condition, are available on the second-hand market for next to nothing, or nothing at all. This is the time for the collector to pounce, as these opportunities seldom last forever. One only has to go back to the mid-1990s, when turntables were unfashionable and even real beauties such as the Beogram 4000 could be bought for very little, something that is certainly not the case any more. If space allows, it even makes sense to store away broken, incomplete or seemingly superfluous examples of a particular model, if it is one you like and available cheaply. When the central stocks run out they will become very useful, as many of the parts used in Bang & Olufsen equipment either defy substitution or are beyond the skills of even the most talented home fabricator.

The collector also has a role in the preservation of old things. It is wise to know which items in your collection have a finite life and which can be used more freely. It is a shame to reduce a once pristine example to worn-out scrap just because it was left switched on all day, every day, often with no-one there to enjoy it. Television sets and compact disc players fit into this category, whereas most radios, amplifiers, receivers and loudspeakers generally do not. Certain models, such as the Beocord 2200 cassette recorder, fall into a strange middle ground. They certainly required periodic maintenance but as the heads are virtually wear-proof, the mechanicals simple and durable, and the electronics free of potentially obsolete (and hence unobtainable) integrated circuits, there is no reason why the machine should not be sustained virtually indefinitely. Recording machines, in general, offer a tantalizing proposition to the collector, for they allow any type of programme material to be enjoyed on a totally authentic system with no 'out of period' extra equipment connected, once the recordings have been made. Players of non-recordable media (turntables and so on) and radio sets require the cooperation of others to enable a particular type of music or programme to be listened to.

No discussion of collecting would be complete without the mention of values. The increase in value of items that accompanies the move of a particular field of collecting from the obscure into the mainstream is a continuous source of discussion amongst collectors. At the same time as they smile, as their once worthless junk regains a monetary value, they complain that the hobby has been ruined as they can no longer find anything for nothing. Electronic collectables of all types are difficult to value accurately and the market for them is generally volatile. Although certain models have raised themselves from being almost worthless to surprisingly expensive, there are no fortunes to be made as things currently stand.

Something of which one can be sure is that the value of most Bang & Olufsen products to the collector is influenced to a great extent by the condition and completeness of the cabinetwork. The beautiful lines are completely spoiled by gnarled woodwork, corroded metal and missing parts. Technical faults of all descriptions can be cured eventually, but severely damaged cabinets will normally relegate all but the very rare to 'spares only' status. Even small parts that are missing will considerably de-value an item. Do not assume that an absent knob or piece of trim will be easily replaceable, unless you have one waiting for you at home. Bang & Olufsen used very few generic parts on the exterior of their equipment and even the smallest control will have been carefully styled to complement the rest of the set's appearance. Technical faults also de-value equipment but the extent to which this effect takes place depends on exactly what is wrong. There is an infinite variety of problems between a defective indicator lamp and clouds of black smoke emerging whenever power is applied; the experienced collector will soon know how difficult common problems are to solve and bid accordingly. The inexperienced must, up to a point, take a chance, pay as little as possible and be prepared to learn from their mistakes.

This book deliberately contains no monetary examples of absolute value for any particular item. Not only would a useful scale of values be next to impossible to compile, it would also represent only a 'snapshot', taken on the day of writing. Instead, next to many of the pictures you will find two ratings, one for availability and one for complexity.

Availability is a simple indication of how likely the collector is to encounter a particular model. Those with five circles are fairly easy to come across, whereas those with only one are quite unusual. From this information it is possible to decide whether it is necessary to tolerate either a high asking price or some drawbacks in terms of condition or completeness. Models that are easily available might not necessarily be cheap; however, you are likely to see many of them, so with experience you can decide if the owner's valuation is realistic.

Complexity is not simply a measure of mechanical intricacy or electronic fiendishness, instead it provides an insight into how problematic a particular model is to operate, maintain and, where necessary, integrate into a system. Generally the more circles (again up to five) the more experience and skill you will need to get the best from the equipment.

It may come as a surprise to the new collector how large the complete B&O range is. It has not been possible to cover every model in detail in this book but they can all be found on the authors' website: www.beocentral.com. Details of the latest new models and the current range can also be found at B&O's own website: www.bang-olufsen.com.

CHAPTER ONE

1925 TO 1964: THE EARLY YEARS

Although outside Denmark, Bang & Olufsen is most commonly associated with its beautifully made and beautifully styled TV and audio products of the 1970s and 1980s, its roots go back much further. Bang & Olufsen is one of the oldest names in the radio industry and enjoys one of the longest records of largely uninterrupted domestic receiver manufacture.

The Bang & Olufsen equipment made up until 1964 is not widely collected. Not only is it scarce outside Denmark, it is also from a period when Bang & Olufsen was mostly just another small radio producer with a range of simple models of broad public appeal. Much of the flavour that characterizes their post-1964 output is absent from this earlier equipment (some notable models excepted) and, as a consequence, most of it is not well known. Serious collectors like to have a few early sets to give some context to the rest of their collection, but for the beginner this should not be considered essential. It is, however, useful and informative to be able to understand how the company came to be founded and how it evolved into the force that it represents today.

The Founders

Unlike the companies of today, whose names are carefully chosen by consultants, committees and market research, the Bang & Olufsen Company takes its name from its two founders: Peter Bang and Svend Olufsen.

Peter Bang was the son of a well-to-do department store manager. He was born on the 14 March 1900 and spent his formative years in Copenhagen, the capital city of Denmark. He acquired an interest in electricity and radio from a young age, and con-

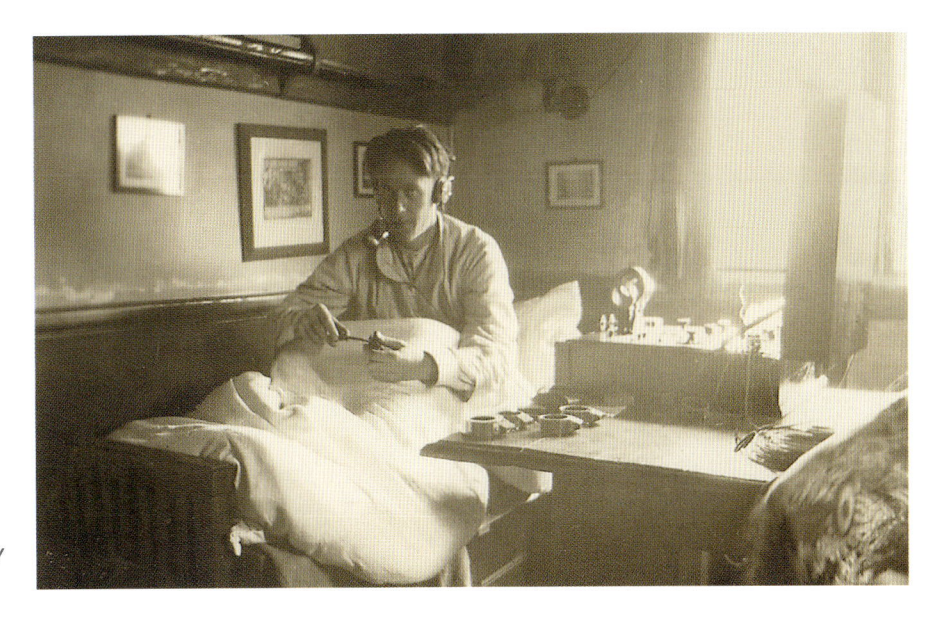

Peter Bang, seen here experimenting with his mains receiver in 1923. *(Picture courtesy of Bang & Olufsen.)*

Svend Olufsen, 1925. *(Picture courtesy of Bang & Olufsen.)*

structed receiving sets and other equipment throughout his time at boarding school and at the Electrotechnical School of Arhus. By 1923 he had constructed a radio set that was powered by mains electricity at a time when most radios needed a number of separate batteries to function.

After graduating in 1924, Peter Bang was drawn to the United States of America, where radio had already started to establish itself as a mass market entertainment medium. He worked for six months both in a radio factory and as a petrol station attendant before realizing that he should start his own business back in Denmark. He predicted that radio would soon be as popular in Denmark as it already was in the United States and that, therefore, opportunities existed for anyone who was able to manufacture receiving equipment.

Svend Olufsen was the son of a Danish parliamentarian. His father owned the Quistrup estate, the site of a former monastery close to the town of Struer, which would later become the famous home of the Bang & Olufsen company. Shy, quiet and dyslexic, Svend Olufsen also had an interest in radio and attended the Arhus Electrotechnical School at the same time as Peter Bang. After graduating also in 1924, Svend Olufsen returned to Quistrup where he continued to experiment with radio sets. Just like Peter Bang, he was keen to develop a practical receiver that could be operated from mains electricity. Taxed by the practical difficulties of constructing the equipment, Svend Olufsen needed a partner to help him in his work. He invited Peter Bang to visit him at Quistrup and a fruitful working relationship was soon formed.

The Bang & Olufsen company came into being on the 17 November 1925. Starting with a capital of 10,000 Danish kroner, the new company started to equip the attic and spare rooms at Quistrup for the production of radio equipment.

The First Products

The first product that the newly founded Bang & Olufsen Company produced was known as the

Eliminator. Its purpose was to eliminate the need for the batteries that powered the radio sets of the period. Similar devices were appearing in other markets at the same time, one of which was produced by E. K. Cole of Westcliff-on-Sea in southern England as the first product of the famous 'EKCO' company.

The design and construction of battery eliminators was not a trivial task. Battery radios needed a high voltage at a low current for the valve anodes, a low voltage at a low current for the valve grids and a low voltage at a high current for the valve heaters. All these supplies had to be smooth and free from noise if clear reception was to be obtained, stable so that the fragile valves of the period were not damaged and safely isolated from the incoming mains network. Radio sets of the mid-1920s had a great number of exposed parts and terminals, meaning that a great risk would be present if the battery eliminator was not designed and produced carefully.

The first mass-produced Bang & Olufsen eliminators appeared in 1926. The growth of both radio and mains electricity ensured their success and soon the business outgrew the accommodation available at Quistrup. A new factory was built nearby to allow the expansion to continue; although it was designed so that it could also be used as a school should radio prove to be a passing fad. New, better and more elaborate eliminators continued to be produced and their popularity was such that one of Denmark's largest battery makers began to feel threatened to the point that they published advertisements warning of the dangers of connecting radio sets to the mains power. Despite this, Bang & Olufsen continued unabated and, by 1928, were the largest manufacturer of battery eliminators in all of Scandinavia.

As well as battery eliminators, Bang & Olufsen also produced complete radio sets. Three- and four-valve models for use on either alternating current (AC) or direct current (DC) mains power were produced, housed in elegant cabinets with no exposed parts. They also made their own moving coil 'electrodynamic' loudspeakers and sound equipment for use in cinemas, which was marketed under the 'Bofa' trademark.

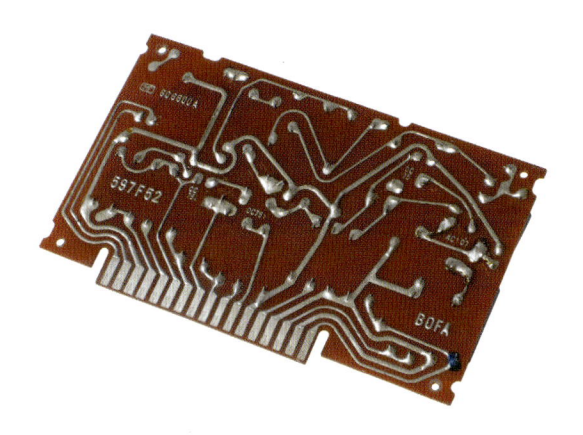

Bofa was a division of Bang & Olufsen that made professional products. They also made parts for other Bang & Olufsen products, like this transistorized amplifier for a tape recorder of the early 1960s.

Their first really noteworthy radio design appeared in 1929. Known simply as the 'Five Lamper' ('five valves'), this set combined all the knowledge that had been gained from the previous work in radio and in cinema sound to produce an attractive set of outstanding performance. Housed in a beautiful cabinet complete with fine wood inlays, this new set could produce a powerful, rich sound. The matching loudspeaker was housed in a separate cabinet and used Bang & Olufsen's own 'type D' drive unit. The Five Lamper was not cheap but it was sufficiently appealing to justify a larger field of distribution than had been attempted up until that time. Bang & Olufsen's fame was beginning to spread because of the quality of their products, something that would be repeated many times in the following years.

The 1930s

The 1930s saw Peter Bang's predicted explosion in the growth of radio come to pass. Broadcasting stations were appearing all over Europe and were providing output of genuine entertainment value, while sets were becoming cheaper, easier to accommodate and easier to use. At the beginning of the decade, the

The B&O bottle opener was first seen in 1937 but has since been re-issued many times.

radio set had already become a single unit with only a couple of controls, powered by mains electricity and capable of running all day without attention. Bang & Olufsen's new models followed this trend and remained popular. Around this time, a new type of instrument became available, the radio-gramophone. These units combined a radio receiver, loudspeaker and turntable unit into one (usually large and imposing) cabinet and were very much at the luxury end of the manufacturers' ranges. Bang & Olufsen produced their first model in 1930, using the Five Lamper chassis and one of their large loudspeakers in an ornate cabinet. The high price initially meant that sales were slow but the possibility of enjoying high-quality music of the listener's own choice at any time, held great appeal and the subsequent models, which could be produced more economically, became very popular.

As the company's fame grew, a more professional attitude was taken with advertising and promotion. 1931 saw the slogan 'Det Danske Kvalitetsmaerke' ('The Danish Hallmark of Quality') coined and during the following year a new trademark appeared. Both are still in use at the time of writing, over three-quarters of a century later. The trademark appeared on both advertising material and, of course, the sets themselves. It was also at around this time that the company began to be more widely known by its initials, B&O. In 1937, the trademark was even adapted into an opener for beer bottles. Even though such trinkets make a nice addition to a collection, it is important not to be taken in. Far from being a rarity, the bottle opener is actually the B&O product of which the most have been made and it has been re-issued many times over the years. Dealers sometimes include one as a gift when a purchase is made, or give one to regular customers. In fact, B&O claim this is the only item they make that can't be bought, as it is always given away free.

Radio technology advanced rapidly during the 1930s and B&O continued to produce sets that were thoroughly up-to-date, in terms of both performance and appearance. A striking set appeared in 1934 that was known as the Hyperbo 5 RG (steel). This large radio-gramophone was housed in a plain black cabinet framed in chromed steel tubes, a style that echoed that of the avant-garde furniture of the day. Though striking today, this design would very probably have been un-saleable at the time, as buyers of such expensive instruments tended to have a very conservative outlook and were more likely to be put off than impressed. As a result, only a handful were made, although the same radio chassis and turntable mechanism were used in a wide range of more conventionally styled Hyperbo models.

1937 saw new ranges of all-wave receivers added to the range that could reliably receive on the short waves, as well as on the traditional long and medium wave 'broadcast' bands. The design of a usable and stable all-wave set relied not only on the new types of valve that were becoming available, but also on the accurate assembly and alignment of the receiver's tuning coils. B&O's solution to this problem was to mount the coils together into a single screened unit, which Svend Olufsen called the 'Radio Heart'. This became a recognizable selling point, even for buyers who were not technically minded and even if, after an initial period of excitement, the short-wavebands were seldom used. Such was the stability of the Radio Heart that when, during the following year, the frequencies of Europe's main broadcasting stations were standardized, it became possible to produce sets that were pre-tuned in the factory to receive certain programmes. Radio-programme selection by pre-tuned pushbuttons would become another B&O hallmark that endures to the present day.

Another important feature of radio design in the 1930s was the use of new plastic materials for cabinet manufacture. Wood had been the traditional material of choice for radio cabinets until this point. It could be formed into attractive cabinets using existing skills and techniques, it was an electrical insulator meaning that the resulting sets were safe to handle and it was acoustically inert so the finished set could have a clean, natural sound. Wood had its disadvantages too. A lot of manual labour and skill was required to cut out the various parts, join them together and finish them appropriately for domestic surroundings.

At a time when the industry was discovering mass production, the hand-crafted cabinet was beginning to look anachronistic, even though it was still what the public demanded. The plastic material most associated with radio sets is Bakelite. Unlike some modern plastics, Bakelite, once formed under massive pressure, retains its shape and finish, even when subjected to heat. It could be coloured during manufacture and could be moulded into complex shapes. Correctly designed parts would come out of the moulds with a deep shiny finish, ready for use without further surface treatment. At last it became possible to produce a complete radio cabinet in a single operation that could easily be mechanized and automated.

Bakelite radio cabinets started to appear in the early 1930s and were usually only seen for basic, budget sets. The image-conscious still demanded real wood finishes, a position that was reversed when collecting old radios became a popular pastime. B&O started manufacturing Bakelite items from 1932 onwards, starting with small parts, such as electronic components, radio knobs and loudspeaker baskets.

Due to the size of the presses needed to form larger items, B&O's first Bakelite radio cabinet did not appear until 1938. Like the rest of the industry, the new material was first used for a small, simple table

Stockpiles of vital materials meant that attractive sets like this Beolit 41 could remain in production even under wartime conditions.
Availability ●
Complexity ●●

set, positioned near the bottom of the range and known as the 'Beolit 39'. Although technically of very limited interest, this set, as well as being B&O's first model with a plastic cabinet, also introduced the prefix 'Beo' to the model range. 'Beolit' was originally a variation on the spelling of 'Bakelite', but later on it would be applied to all small radio sets, regardless of what cabinet material was used.

Activities During the War

The Second World War came to Denmark in April 1940 when Germany occupied the country. There had been a grim inevitability about this event and so preparations had been made at B&O, with such essentials as radio valves and the raw materials necessary for Bakelite manufacture being stockpiled against the shortages that may come. The industrial capacity of Denmark was not as heavily hit during the war as that of England and, as a result, it was possible for B&O to continue to produce a full range of sets, from cheap small table radios to large high-quality radio-gramophones.

The radio-gramophone line was given a technological boost in 1943, when an advanced automatic record changer was designed. This device could automatically play a stack of discs of mixed sizes without the listener having to intervene in any way and was such an impressive feat of engineering that advertising copy even went as far as to call it 'The thinking record changer'.

The range and quality of B&O's output during the war years would have been the envy of the British listener, who on the whole had to rely, during this time, on ailing pre-war models (for which certain key replacement parts were not available) or cheap, poor-quality miniature American sets.

It was inevitable that, as a collection of skilled and ingenious people, members of the company's workforce would become involved in the resistance to the occupation. The company as a whole had refused to comply with the occupier's demands, a fact that made them stand out amongst the larger Danish radio manufacturers. The most important piece of resistance work that was carried out was done by Lorenz Duus Hansen, holder of the top position at B&O's office in Copenhagen. Radio communication was of vital importance to the resistance movement but the sets that they had been sent from the UK and other sources could not operate on the direct current mains supply that was still in common use in parts of Denmark at the time. Duus Hansen designed a miniature transmitter and receiver set that could run on Danish mains power and could fit into a briefcase, meaning that it could be transported and hidden relatively easily. This set became known as 'The Telephone Book' because it was similar in size to a Danish telephone directory. Although the set was not produced in B&O's factory, Duus Hansen's connections to the company were well known. When it was discovered by the German army that Duus Hansen was still being paid by B&O despite being in hiding, it was only a matter of time before revenge was taken.

On the night of the 14 January 1945, just over four months before Denmark was liberated from German control, the B&O factory was completely destroyed by a bomb planted by a team of Danes under the command of German counter-sabotage. The devastation was total and, of course, stopped production at a stroke. Despite no longer having its main production facility, B&O did not make any of their employees redundant, even though full-scale production would not resume for another two years.

Post-War Recovery

As the production facilities were gradually rebuilt, B&O were once again able to resume their manufacturing activities. Shortages of parts for radio sets brought a brief diversion into electric razors, a little known B&O sideline that continued well into the 1950s. At the other end of the complexity spectrum, 1947 also saw B&O launch the Beocord 84U, Europe's first magnetic recording machine to be available for domestic use. The origins of the Beocord 84U came from an American military machine that was given to

1925 TO 1964: THE EARLY YEARS 15

the wire recorder was also built into some of B&O's top-line radio-gramophones. The mechanism for the recorder was integrated with that of the record player, the turntable rim doubling as one of the spools on which the recording wire was stored. Models including this highly unusual arrangement first appeared in 1948 and by 1950, a portable model had appeared that used a similar mechanism. By 1951, steel wire had been replaced by plastic tape in a new range of Beocords. Combination tape recorder/record player models remained in the range and, despite their odd appearance and their being slightly odd to operate, they remained popular.

There was no point in developing recorders if there were no microphones to go with them. The original models were very basic but standards soon improved with the adoption of the 'ribbon' principle. Types such as the BM 3 and BM 5 were expensive when new but were well-respected and are very collectable today.

A BM 5 ribbon microphone, in its mono configuration.

Television

Television came quite late to Denmark, so B&O had little or no involvement with it in the pre-war years. The television systems that had been established before the Second World War in a number of countries had shown the direction the future would take, and had made it clear that there would be no room in the mass market for manufacturers that relied on radio alone. As a company who partly depended on equipping cinemas for their continued financial well-being, B&O were surprisingly keen to embrace television and produced a small number of prototypes that were first exhibited in 1950. These sets were completely engineered by B&O themselves and represented a considerable achievement for a small company based in a country with no television infrastructure of its own.

The first range of B&O television sets became available in 1952. These sets were given model numbers similar to those used for radios; the name 'Beovision' was not used until 1964. The early sets were large and cumbersome. One model included pull-out handles and small wheels so that it could be moved easily,

Lorenz Duus Hansen during the war as part of a system that could make 'high-speed transmissions'. This was desirable, as reducing the time taken to make a secret broadcast considerably reduced the chances of detection by enemy forces. Immediately following the war, B&O were able to secure the right to use the AC-bias recording system that the American machine used and that was responsible for its music-grade performance. The Beocord 84U did not use plastic or paper tape coated in iron oxide; instead it used steel wire, which, although not capable of the same levels of performance as the best magnetic tapes, was easier to produce at the time.

As well as being available as a separate instrument,

earning it the nickname 'the wheelbarrow'. There were no Danish television broadcasts in 1952 but viewers in the southern parts of Denmark could receive German programmes from across the border. As transmitters and receivers improved, these programmes eventually became receivable as far north as Struer, covering so much of the country that the Danish Government were forced to act and to start the process that would finally give Denmark a television network of its own. This resulted in a 'TV boom' around 1956, during which B&O did very well, selling their 'Capri' and 'Grand Prix' models in large quantities.

Stereo Sound

The excitement that television brought overshadowed radio in the public's mind. Radio sets were no longer the only source of electronic entertainment in the home and it was clear that the market for large table sets would soon decline. To arrest this change, the industry as a whole re-positioned the top end of the radio line into a new field known as 'high fidelity'. The aim of high fidelity was to provide musical reproduction that was as accurate as was technically possible, a field that was given a major boost in the late 1950s, when stereo recordings arrived. B&O embraced stereo wholeheartedly, introducing both stereo tape recorders and a stereo record player based on a magnetic pickup of their own design. The pickup, known as the SP 1, was of a striking design, small and tapering to a point from where the stylus emerged. It used a magnetic principle that would endure in a miniaturized form until B&O ceased the production of record players in the 1990s. Along with the pickup, a range of new turntables was developed, lighter and smaller than those that had come before. The B&O stereo turntable of 1960 was a great success and, in various modified forms, remained in the range until the early 1970s. The SP series magnetic pickup produced only a very small output signal when compared to other makers' ceramic and crystal offerings, so the B&O turntable could be ordered with a special

An early B&O SP magnetic pickup.

amplifier built in to boost the signal to a more conventional level.

Becoming a force in high fidelity audio proved to be a wise move by B&O. When the television boom died down it left them with new markets on which to base further successes. As trading restrictions were dismantled across Europe in the following years it was with stereo audio products that B&O led the export drive that ensured their future survival.

A New Look for a New Decade

As the 1950s drew to a close, the mood of post-war austerity was finally beginning to lift across Europe, to be replaced with one of confidence and optimism.

The effects of this can be seen in the products of the period, which were characterized by bold styling and bold colours. Even in conservative Denmark, B&O realized that the market for large radio sets in the traditional style would soon become limited, as the buying public were beginning to demand something new. It was, however, important not to alienate the more conservative customer, so an interesting policy towards cabinet design was developed. Technically identical radio and TV models would be offered in very different cabinet styles, some with traditional fine veneer finishes, rounded corners and muted colours, and some in a more 'architectural' style, boxy with flat surfaces, unfinished wood and bright painted details. The ultimate expression of the latter style came in the form of a 'modular' system, where units such as a TV set, a loudspeaker, a shelf unit or a radio receiver were all made in plain rectangular cabinets that could be stacked together on a special table to form a system as simple or as complex as the owner required. A range of finishes was offered, from hardwood veneers to stripped pine, to bright painted metal in contrasting colours. The appearance of this equipment was very typical of the Scandinavian style of the period, which no doubt impressed a section of the home market but did not do much for exports. Eventually the new look was accepted, the older cabinet styles disappeared from the range and B&O were on the way to becoming the leaders in audio-visual styling, as they fully established themselves during the 1970s. The 'Dirigent' FM tuner-amplifier of 1962 showed an early attempt at the long, slim look that would become the most readily identifiable B&O hallmark, although the extent to which this look could be applied was limited by the hot valves and bulky components that all electronic equipment of the period still used.

The boxy shape of this Mini 610 De Luxe radio was a brave styling exercise in 1962, as it was not what traditional radio buyers expected.
Availability ●●
Complexity ●●●

Transistors – a New Role for Radio

During the 1950s, television had marginalized radio as the main domestic entertainment medium. While big table radio sets were to have a limited future, television had not yet proved a match for the convenient portable radio set that could be used anywhere, indoors or out. Portable radios improved greatly with the invention of the transistor. Portable valve sets had been available but they required large, expensive batteries and were capable of only limited performance. Transistor sets, although initially more expensive to buy, could be operated far more economically, were more robust and, because of the small size of the components and batteries they used, could be styled very attractively.

B&O were not the first to produce a transistor radio but by releasing their first model in 1959, they entered the field promptly for a comparatively small company. In the following year they introduced a further model that covered the FM band, some years before such a set would appear in any British manufacturer's ranges. All B&O transistor radios would use the name 'Beolit', even though none of them had Bakelite cabinets.

Beolit portable transistor radios were expensive but capable of excellent sound quality. They operated using torch batteries, rather than special (and expensive) radio types and so, even now, can be operated economically. They are a consistent favourite with collectors, as they do not take up too much space and are not very difficult to repair. Not all models were offered in all markets but, as portable radios were designed to travel, they can turn up anywhere, if one keeps one's eyes open.

Transistors would be very important in B&O's product ranges of the following decade. They would pioneer transistorized radios, tape recorders and high-fidelity equipment, but they would be amongst

The original Beolit transistor radio of 1959.
Availability ●●
Complexity ●●

RIGHT: The AM-only version of the Beolit Teena transistor portable.
Availability ●●
Complexity ●

LEFT: A close-up view of the tuning dial of the AM Beolit Teena.

RIGHT: The Beolit Teena was also available as an AM/FM set.
Availability ●●
Complexity ●

20　1925 TO 1964: THE EARLY YEARS

LEFT: The 1960s Beolit 600 was part of a large series of similar models.

BELOW: The clear dial was marked with station names as well as frequencies.

LEFT: This miniature meter showed both battery condition and signal strength.

the last of the major manufacturers to produce a fully transistorized colour television set. Early transistors were expensive and of limited capabilities, so B&O, like many other firms, mixed them with valves in some models to obtain the best balance of reliability, performance, size and value in a particular model. This 'hybrid' technique was most popular in televisions but it appeared in other lines too; for example, the Beocord Belcanto tape recorder of 1963 featured an amplifier that used both transistors and valves. This model used a tape deck made by BSR of Birmingham that, elsewhere, was commonly fitted to budget recorders, a fact that can surprise those who are more familiar with the elaborate semi-professional models at the other end of the Beocord range.

1925 TO 1964: THE EARLY YEARS   21

RIGHT: **The Beocord Belcanto. The BSR TD2 tape deck it used was also a familiar fitment in British budget machines, though B&O fitted better heads to improve the performance.**

This view gives a better impression of how the rather unusual folding cabinet is arranged.

In common with many 1960s designs, the Beocord Belcanto used a mixture of valves and transistors. Two of the valves can be seen towards the centre of the picture beneath the loudspeaker, whereas the four transistors are mounted on the printed circuit board on the left. Further components are mounted below the tape deck.

CHAPTER TWO

1964 TO 1969: NEW MARKETS, NEW TECHNOLOGIES

The removal of barriers to trade across Europe in the mid-1960s posed a particular problem for small manufacturing companies like B&O. No longer could they rely on protection from imported equipment by tight quotas and heavy taxes – a factor that meant that their local market share would inevitably fall. This period saw many once-famous names either disappear or be swallowed up by the giants of the industry. Philips bought up many ailing concerns in the years that followed. In Denmark, the radio manufacturer Arena, a firm that had been a rival to B&O, was acquired by the Rank organization and soon became merely a name that was applied to Rank's standard range of products. As well as competition from inside Europe, B&O also had to take on the Japanese, whose industry was growing at lightning speed and was beginning to produce some products of real quality. Difficult times lay ahead and some new thinking would be required if the company was to survive.

The choice that B&O made was to change direction completely. Instead of offering a complete range of radio and TV models, they would concentrate on the upper end of the market, offering superior styling and advanced technology in return for a higher

This Beolit 700 is a typical 1960s B&O product, well made and capable of excellent performance.
Availability ●●●
Complexity ●●

A close-up view of the tuning dial of the Beolit 700 shows the two separate scales: one for FM and the other for the three AM bands. The instruction book for this model described how the set could be used for direction finding as well as normal reception.

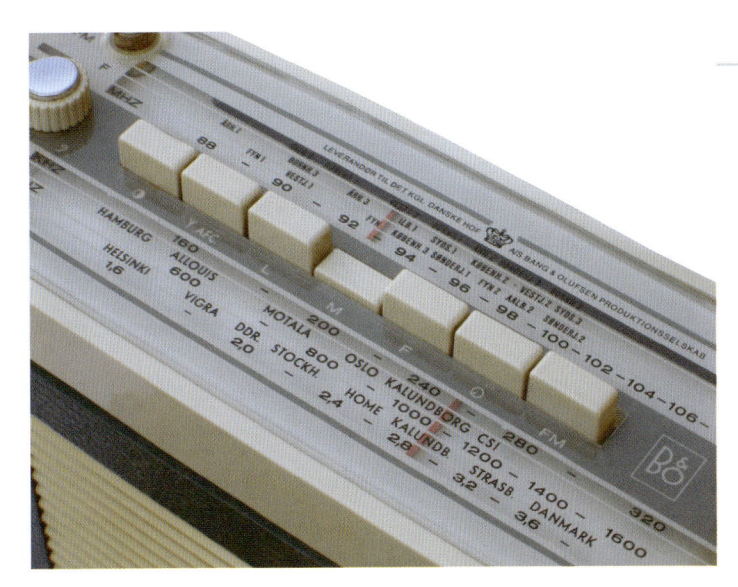

A similar chassis was also used in this model, the Beolit 800. The sturdy wooden cabinet slightly improved the sound quality.
Availability ●●●
Complexity ●●

asking price. They would also no longer focus completely on the Danish market but, instead, produce designs with real international appeal that could be exported in quantity, therefore massively enlarging their market and earning much needed foreign currency. This new ethos could be summarized by the popular marketing slogan of the period that was translated in various forms, one of which was:

'Bang & Olufsen – for those that consider design and quality before price.'

These new plans were not just left as good intentions, they were backed up with a range of exciting new models, which fully met the requirements of the new policy. Some of the most important ones are detailed below. Most B&O collectors consider this new range as their starting point, not only because

The Beolit 800 could be used like this as well. Lowering the handle caused two small feet to rise from the front of the cabinet, protecting the woodwork.

these products are particularly interesting but also because they were widely exported and so it is easier to find good quality examples without having to travel great distances, or arrange international shipping or convert equipment intended for other electrical or transmission standards.

Beomaster 900

If one product epitomized B&O's transformation from a small radio manufacturer, serving a naturally loyal national market, to the world-famous marque known for innovation, style and quality wherever it is represented, then that product must be the Beomaster 900. Launched in 1964, this small table set caused a sensation. For its size it was capable of truly remarkable performance and yet remained elegant and easy to operate. Undoubtedly the most desirable new model of the season, the Beomaster 900 became an international hit and spearheaded export drives into new markets.

What made the Beomaster 900 special was that the designers had discarded valves in favour of transistors. Transistor radios were not new, they had first become available in the mid-1950s; B&O had produced their first transistorized set in 1959. However, these first sets were portable models, battery operated and intended for basic reception of local stations. What made the Beomaster 900 novel was that it was not a portable set but a fully-featured table model aimed at the discerning listener. Transistor radios had not been offered in this market sector before because valves were still the dominant technology. Using transistors gave the designers of the Beomaster 900 certain freedoms that were not available to those still using valves. Instead of attempting to make the Beomaster 900 very small, B&O left the size very similar to the previous year's Mini De Luxe but instead opted to cram it with every feature imaginable. The most important change was that a stereo amplifier was fitted; compact, cool-running transistors making this practical, even in a small cabinet. At 6W per channel, the amplifier was considered powerful at the time, offering as much power as a radiogram or a small hi-fi setup. The stereo amplifier

1964 TO 1969: NEW MARKETS, NEW TECHNOLOGIES 25

The Beomaster 900K was the first transistorized table radio. It sold in huge numbers and is still a practical receiver today.
Availability ●●●●
Complexity ●●

could be used in conjunction with a stereo record player or stereo tape recorder and, if the owner wished, an FM stereo radio decoder could be fitted at extra cost. The radio covered four wavebands including the 'international FM' band right up to 108MHz and came complete with automatic frequency control (AFC) for drift-free FM reception, separate pointer drives for AM and FM tuning, and an illuminated signal strength meter.

The most popular variant was the 900K. Styled by Henning Moldenhawer, this version had two built-in loudspeakers, one at each end of the dial. The use of transistors in the main chassis gave an unexpected advantage to the performance of the loudspeakers. Previous sets had needed a free flow of air over the hot valves and power supply components, which was usually achieved by making large numbers of holes in the rear of the cabinet. This in turn meant that the loudspeaker, if it were to be fitted in the same space, was by definition operated in an 'open backed' configuration. Better performance could be obtained from a loudspeaker if its rear could be sealed in an airtight box, the size of which determined the available bass response. The cool-running Beomaster 900 chassis could be easily sealed in an enclosed box along with the loudspeakers, but without some clever thinking a very large cabinet would have been necessary in order to produce a full-toned sound. In the Beomaster 900K, however, the internal space was sub-divided into three areas: one for each loudspeaker and one for the radio chassis. These areas were joined together by internal ports, which were sized in such a way that the effective volume of each loudspeaker

The green stereo indicator would illuminate when an FM stereo broadcast was being received, if the optional stereo decoder was fitted.

cabinet was equal to its own area plus the area occupied by the chassis. This gave the Beomaster 900K full 'big set' performance, despite its small size. The sealing of the cabinet was so effective that labels had to be pasted inside warning anyone working on the set not to play it at high volumes with the chassis out of the cabinet, since the loudspeaker units, unrestrained by air pressure, could be damaged.

Other variants included the 900M receiver, which had no built-in loudspeakers, and the mono Beomaster 700, which had only one. In addition, various large radiogram cabinets were also offered, complete with either a B&O manual record deck or a fully automatic Garrard record changer. Both decks came fitted with a B&O SP series magnetic pickup cartridge that could be considered very advanced for a radiogram. A space was left in the cabinet to house a B&O stereo tape recorder but, as this was very expensive, the same space could also be used to store LP records instead.

The Beomaster 900 line was later augmented by the similarly sized Beomaster 1400. This model used silicon transistors and was considerably more powerful. Both models were replaced in 1970 by the Beomaster 1600, the final model in this series.

The Beomaster 900M was designed to be used with external loudspeakers.

Henning Moldenhawer styled the Beomaster 900 replacement, the Beomaster 1400. Similarities to the television styling of the day can be seen in this picture, most notably the black grille at the rear and the aluminium framing of the black front panel.
Availability ●●●
Complexity ●●

1964 TO 1969: NEW MARKETS, NEW TECHNOLOGIES

All Beomaster 900s are very collectable and make excellent radio sets for everyday use. Such is their quality and reliability that it is not unreasonable to hold out for a fully working example, although a clean and tidy set with minor faults is also worthy of interest. It is nice to find one with the optional stereo decoder fitted, but bear in mind that an outdoor antenna is desirable if you intend to receive stereo broadcasts, as sensitivity is generally poor.

Stereo Tape Recorders

Not content with just launching the Beomaster 900, in the same year, B&O also introduced a range of fully transistorized stereo tape recorders. These impressive machines were based around B&O's own mechanicals, along with modular electronics that allowed the user to tailor the machine to their own requirements. Three heads were fitted, which allowed the recording to be checked whilst it was being made, a feature that would normally be found only on very expensive studio-based recorders. The initial model was known as the Beocord Stereomaster, although design evolved with numerous detail improvements, until a few years later the Beocord 2000 De Luxe emerged. This was a very fine machine indeed and included a built-in stereo mixing desk with separate left and right faders for each input source. This, combined with a recorder that was capable of professional levels of sound quality, made the Beocord 2000 De Luxe effectively a recording studio in a box. Two versions were sold: the K model, which had a wooden cabinet and a clear plastic cover; and the T model, which came in a robust luggage type case. The T model came complete with a pair of stereo loudspeakers built into the lid. These recorders were amongst the most expensive items in the B&O range but their high quality and versatility meant that they sold strongly throughout a long production life.

For those who wanted to make high-quality recordings but did not need the complex mixing facilities, the Beocord 1500 was later introduced. This model featured the same mechanicals as the 2000 range but had simpler controls. It was ideal for connection to an existing stereo system, such as the Beomaster 900.

For buyers on even tighter budgets, a series of basic mono versions was also introduced. The best-known

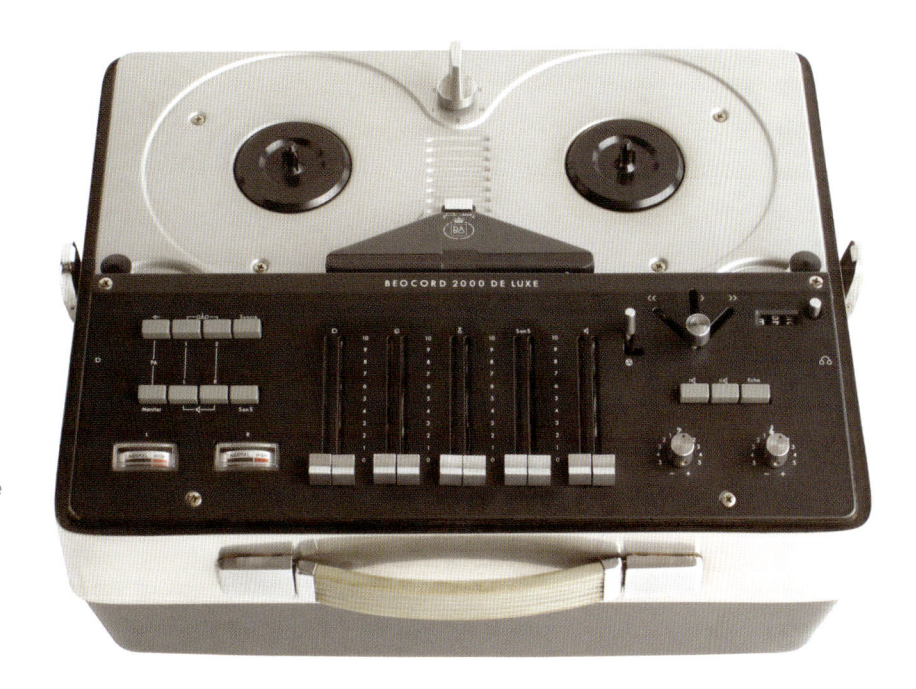

The Beocord 2000 De Luxe was one of the best tape recorders of the 1960s. It offered the facilities of a small recording studio in a neat, convenient package.
Availability ●●
Complexity ●●●●

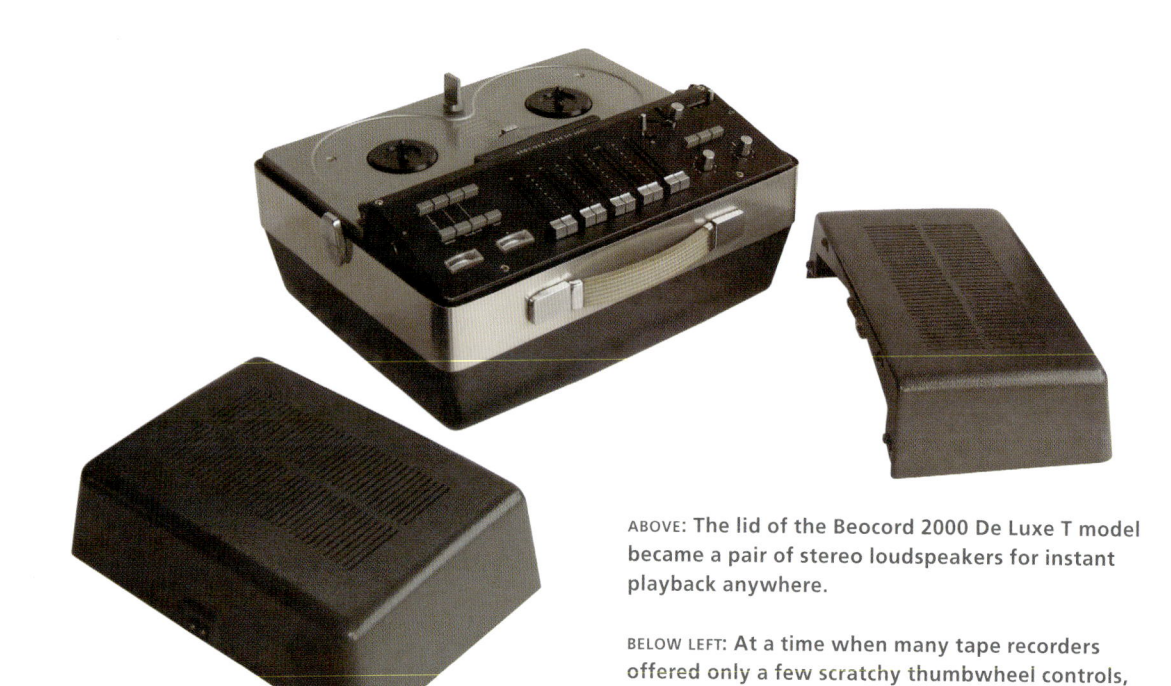

1964 TO 1969: NEW MARKETS, NEW TECHNOLOGIES

ABOVE: The lid of the Beocord 2000 De Luxe T model became a pair of stereo loudspeakers for instant playback anywhere.

BELOW LEFT: At a time when many tape recorders offered only a few scratchy thumbwheel controls, the Beocord 2000 De Luxe included a proper stereo mixing desk.

BELOW: This view inside one of the loudspeaker units shows the unusually slim drive unit and the specially moulded spaces for storing microphones and cables.

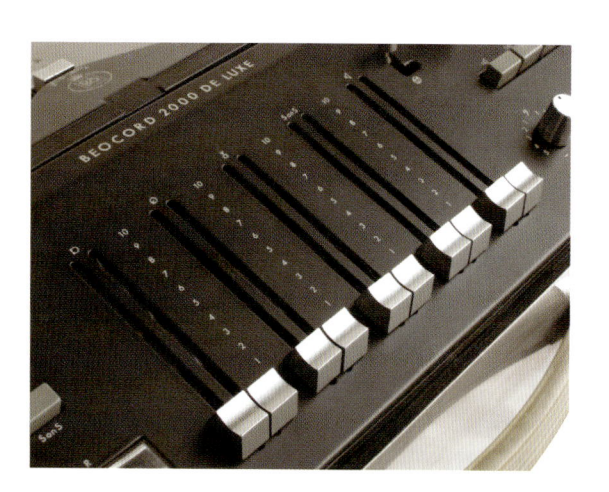

BELOW: The BM 5 stereo ribbon microphone was ideal for use with the Beocord 2000 De Luxe.

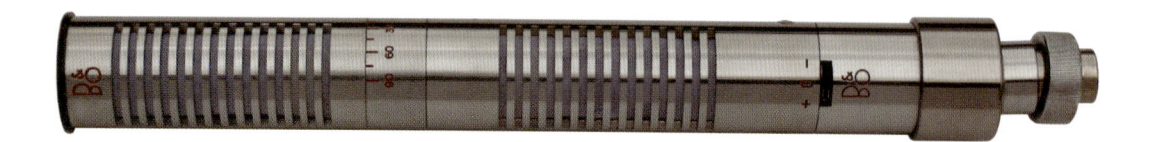

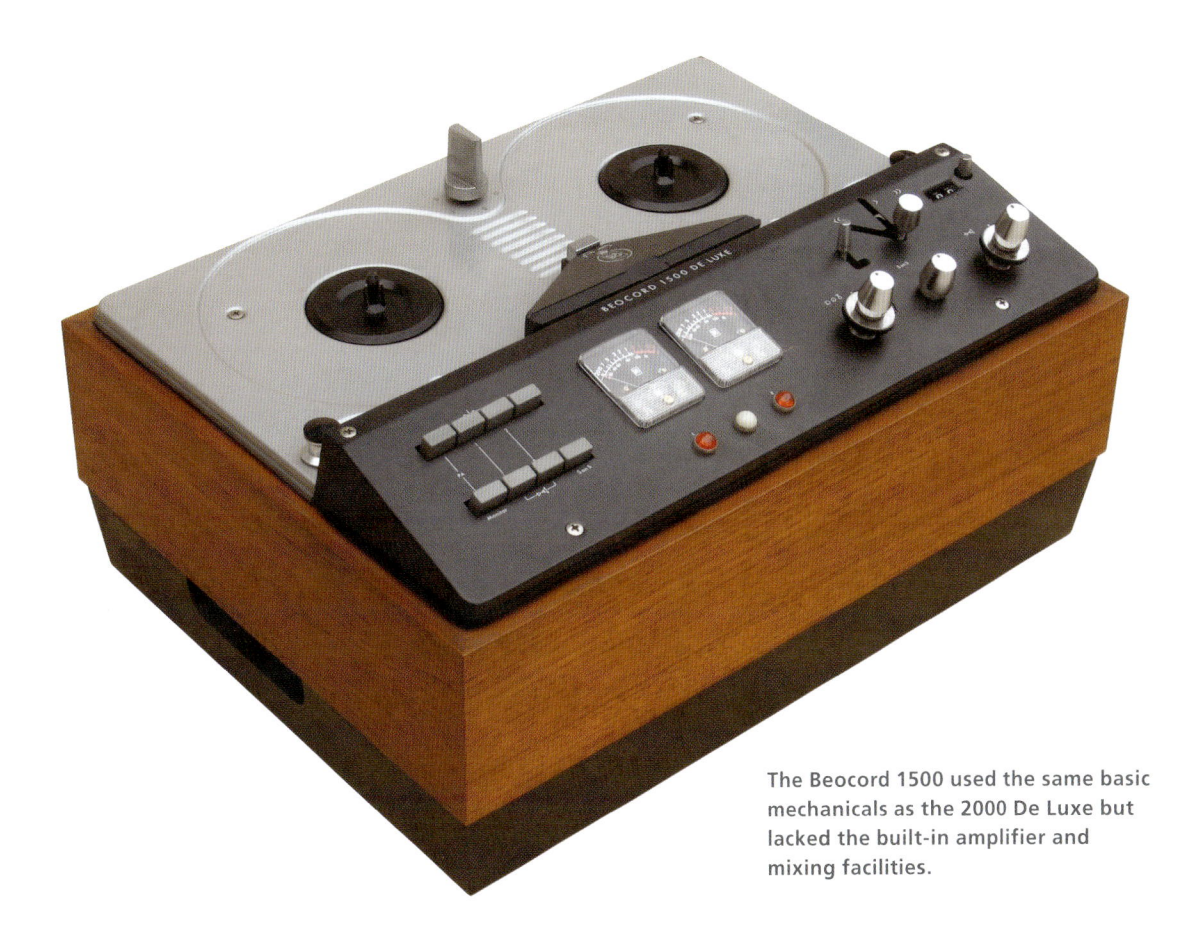

The Beocord 1500 used the same basic mechanicals as the 2000 De Luxe but lacked the built-in amplifier and mixing facilities.

of these was the Beocord 1100 and, unlike the other models, it featured an automatic level control and a built-in loudspeaker.

When inspecting a Beocord 2000, ensure that all the tape heads are in good condition, as the windings tend to fail with age and obtaining replacements is difficult. These machines can still make excellent recordings, if quality tapes are used, and their mechanical sections can be easily overhauled, if care is taken.

Beomaster 1000

The Beomaster 1000 was a transistorized tuner amplifier that first became available in 1965. It was intended to be the main component in a home audio system, so two record players and a tape recorder could be connected, along with two pairs of loudspeakers. The B&O stereo record player was renamed 'Beogram 1000' to show clearly that it was a natural part of the system.

The Beomaster 1000 was capable of producing 15W per channel, making it much more powerful than the Beomaster 900. It also met the DIN standard requirements for hi-fi quality. The amplifier was sufficiently powerful for a warning to be printed on the rear panel that stated that it 'may destroy small loudspeakers'. The tuner covered the FM band only and could be fitted with a stereo decoder at extra cost. The main controls were in the form of keys, neatly distributed with five on each side of the tuning dial. Both the keys and the dial were marked on their front and top faces so that the equipment could still be operated easily if placed on a high shelf.

1964 TO 1969: NEW MARKETS, NEW TECHNOLOGIES

ABOVE: The Beomaster 1000 was kept right up-to-date with a new black finish and a redesigned circuit using the latest silicon transistors.
Availability ●●●
Complexity ●●

RIGHT: This close-up view of the control keys of the Beomaster 1000 also shows the small meter, which showed that the tuning was correctly centred on the station being received.

BELOW: The Beomaster 1000 was unusually powerful for the time, as this warning printed on the rear panel demonstrates.

1964 TO 1969: NEW MARKETS, NEW TECHNOLOGIES   31

TOP: A Beogram 1000 like this one is ideal for use with many B&O models of this period, including the Beomaster 1000.
Availability ●●●●
Complexity ●●

RIGHT: One way to complete the system was with these attractive Beovox 1600 loudspeakers, which could be hung on a wall like pictures.

The Beomaster 1000 was completely revised in 1967 to make use of silicon transistors and, although the basic appearance stayed the same, most of the inside was new. The new version could easily be recognized as the keys and dial surround were made black instead of white, and the whole unit was slightly less fussily styled and labelled than the original. This piece of work is said to be the first project for B&O undertaken by the designer Jacob Jensen, whose later work resulted in the long, sleek models that would characterize the B&O products of the next generation.

Beolit 500

Having already demonstrated with the Beomaster 900 that transistor radios need not necessarily follow the portable format, in 1965 B&O launched another startling new concept. The Beolit 500 was styled in such a way that it was more at home on a desk or hung on a wall than it was on a kitchen window sill or out at a picnic. The new styling was not the only novel feature. The Beolit 500 was an FM-only radio, where the tuning was pre-set. Programme selection was performed as it is with a television – five numbered keys gave instant access to the stations. The only other obvious control was for volume – the pre-set tuning controls and the tone control were hidden away underneath. The Beolit 500 was fitted with automatic frequency control (AFC), so once the stations were set, they could not drift off tune.

Removing the tuning dial freed the external appearance of the Beolit 500 from a lot of the usual 'clutter' associated with radios, making it look sleek and modern. The pre-set tuning function was implemented using a fully electronic tuner that completely dispensed with mechanical variable capacitors and drive cords. This was a real step forward and would soon become universal in B&O FM radio design. The Beolit 500 was battery operated (a particular advantage for a wall-mounted set, since then there were no cables to hide) and needed, as well as the regular torch-type batteries for the radio, a special 22.5V battery for the electronic tuner. Very little power was needed from this battery and it lasted so long it was often forgotten about altogether.

The Beolit 500 could also be used as an amplifier for a record player and as a source for a tape recorder.

RIGHT: **The FM pre-set electronic tuner of the Beolit 500 meant that there was no need for a tuning dial. This gave the set a unique and modern appearance.**
Availability ●●
Complexity ●●●

LEFT: **The tuning controls of the Beolit 500 can be seen in this underside view.**

1964 TO 1969: NEW MARKETS, NEW TECHNOLOGIES   33

The Beolit 1000 was B&O's most advanced portable radio. They are very collectable; this one is especially desirable because the cabinet sides have the optional real leather finish.
Availability ●●
Complexity ●●

This was not uncommon, but the Beolit 500 had one more function that made it truly unique. By connecting an extension loudspeaker, it could be used as a two-way intercom. Both loudspeakers were also used as microphones for this feature and very good sound quality was possible.

As with all portable radio sets, cabinet condition is everything with the Beolit 500. Make sure that it is complete and that the battery compartment is not corroded. The 22.5V tuning battery is long obsolete but a replacement can be made using two 12V miniature cells of the type that is intended for car alarm fobs and this will give many years of trouble-free use.

Beolit 1000

The Beolit 1000 of 1967 took the development of the portable transistor radio as far as the technology of the day would allow. It was a fully featured set that lacked nothing. It covered all four wavebands, including FM, offered band-spread tuning complete with a fine-tune 'expansion' function on SW and was fitted with a powerful amplifier and high-quality loudspeaker. The FM band could be tuned freely in the normal manner and, in addition, there were three pre-set stations. The FM tuner used a similar all-electronic design to that of the Beolit 500 but, instead of needing a special battery to power it, an electronic DC–DC converter was used to step up the voltage of the normal cells. The amplifier included controls for treble and bass, and could be used with an external source, such as a record player or a tape recorder.

The Beolit 1000 was designed to be versatile. As well as being a high-quality portable radio, it could also be used as a car radio or as the centrepiece for a basic home audio system. When running from its internal batteries, the amplifier could produce 1W, a healthy figure for a portable. However, when an external power supply and loudspeaker were connected, the maximum power was automatically raised to 7.5W – a very substantial amount of power for such a small set. A car-mounting bracket was available, which could be adapted to any type of car electrical system. The standard finish for the front and rear cabinet panels was teak or rosewood but at extra cost one could specify real goatskin leather instead. As the biggest and most powerful of all the Beolit models, the Beolit 1000 is always in demand; tidy, working examples can be very expensive. A damaged example can be bought for considerably less but restoration is likely to be an expensive or complicated exercise.

High Fidelity

Although well-designed and soundly constructed, none of B&O's range of radio products could yet be described as 'high fidelity'; that is, capable of producing the very highest quality sound that is technically possible. The decline in the side of the business that dealt with producing equipment for cinemas (a factor brought on by the spread of television) meant that B&O had to consider new areas of activity. To use as much of the experience that they had gained from cinema as possible, they chose to construct Europe's first true high-fidelity system.

Announced in 1967 and known as the Beolab 5000, the high-fidelity system comprised a number of separate components. Most important of these were the Beolab 5000 amplifier and the Beomaster 5000 FM tuner. Both of these units were constructed to professional standards of performance and reliability but were styled and finished to be acceptable in any domestic environment. Such equipment had not been produced before, and the Beolab 5000 quickly won awards for its industrial design.

The performance of the main Beolab 5000 components is so perfect that they still stand the closest scrutiny today. The amplifier, rated at 60W per

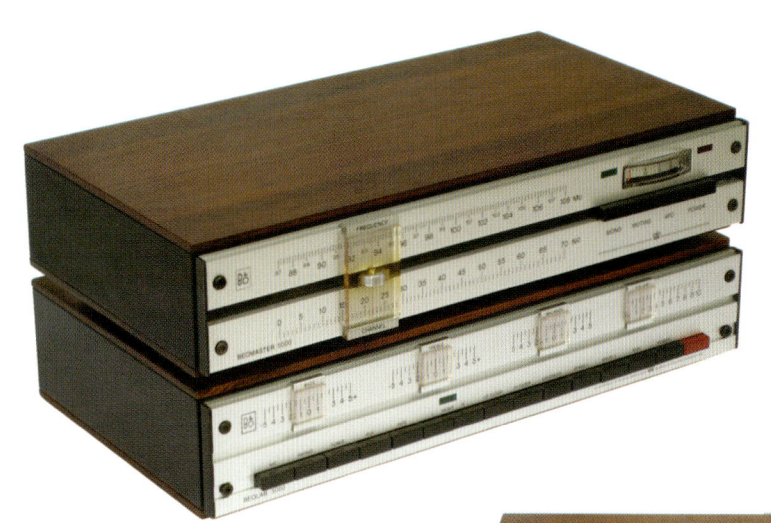

LEFT: **The Beolab 5000 amplifier and the Beomaster 5000 tuner formed the centrepiece of the Beolab 5000 system.**
Availability ●●
Complexity ●●●

RIGHT: **The attention to detail continued even at the back. Note how the tuner is powered from an outlet fitted to the amplifier and that its casework is shaped to allow the free passage of air over the heat sinks, even when the units are stacked.**

1964 TO 1969: NEW MARKETS, NEW TECHNOLOGIES   **35**

The tuning dial of the Beomaster 5000 tuner looked just like a slide rule and so discretely communicated precision and accuracy.

These recessed cap-head screws discreetly recalled the rack-mount handles of studio equipment.

channel (or 120W if used for mono reproduction using a single loudspeaker), was a big step forward in quality and power. It was many times more powerful than any other domestic equipment available at the time and offered a new musical experience for those lucky enough to be able to afford it. The tuner was also of the most up-to-date design and included ceramic IF filters and an integrated stereo decoder.

As well as making the units small enough to fit into a home interior, one key styling challenge was to arrange the complex controls so that they would be attractive to look at and easy to operate. The designer, Jacob Jensen, solved this problem by making the

The complete Beolab 5000 system.

major controls work like slide rules. This not only made them easy to read and use, it also conveyed a subtle message of mathematical precision that fully complemented the accurate, finely tuned sound that the system could produce. The striking appearance of the Beolab 5000 amplifier and the Beomaster 5000 tuner made them perfect as the centrepiece of a key scene of Stanley Kubrick's film *A Clockwork Orange*. This work was set some time in the future at the time of making, an illusion that was helped in a small way by B&O's timeless styling.

As well as the two main units, the Beolab 5000 system also included a choice of loudspeakers. The most suitable of these was the Beovox 5000, a mighty floor-standing design that used multiple drive units to cover the full frequency range.

One problem with loudspeakers of this period was that they did not disperse high-frequency sound very evenly; the listener had to position themselves carefully if they wished to hear the complete range of tones at the correct intensity. A solution to this problem was offered for Beolab 5000 owners in the form of the Beovox 2500 'cube', a small treble radiator that used six small cone loudspeakers to spread the sound in all directions. Although attractively styled, the Beovox 2500 was not the complete answer and was soon made obsolete by the improved dome tweeters that were fitted to later versions of the Beovox 5000, along with most of the other loudspeakers in the B&O range.

The Beocord 2000 stereo tape recorder was a natural partner for the Beolab high-fidelity system, but

1964 TO 1969: NEW MARKETS, NEW TECHNOLOGIES   37

there was no suitable record player in the range that performed to a high enough standard. This problem was temporarily solved by importing turntable mechanisms from Thorens and Acoustical (a little-known Dutch manufacturer) and fitting them into B&O plinths and with B&O ST-P arms and SP pickups. Confusingly, both models were known as the Beogram 3000; further confusion would be caused later as the same model designation was revived in the 1970s and the 1980s for two more completely different models.

A complete and correctly functioning Beolab system is one of those things that every B&O collector would like to own. The tuner and amplifier are the most popular components, only the real purist insists on having the correct loudspeakers (which are very large) as well. Remember that the Beovox 2500 units are designed to work together with these and are not much use on their own or with other, more modern, loudspeakers. Despite not being strictly correct, some collectors prefer to use the later Beogram 4000 turntable with their Beolab 5000. This forms a very nice system, especially when teamed with some later Beovox loudspeakers.

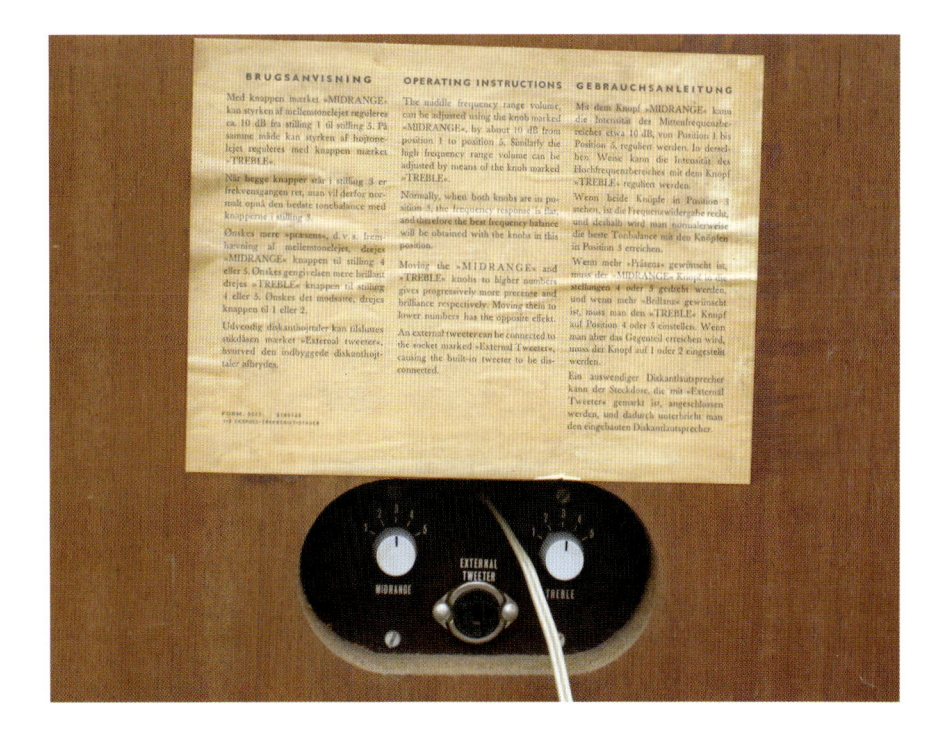

Although stylish, these Beovox 2500 cube loudspeakers are only useful if you also have the large Beovox main loudspeakers that they are designed to work with.
Availability ●
Complexity ●●

This rear view of a Beovox 5000 loudspeaker shows where the Beovox 2500 would be connected. The control knobs on either side are used to balance the tone of the sound.

The printed numbers on the front panels of the Beolab components are fragile and are often found to be worn away. A professional engraver may be able to restore them, if this is found to be the case; though it is important to realize that the scale on the Beomaster 5000 tuner is not linear, so do not clean the old numbering off before noting precisely the location of each digit. Buying the tuner and the amplifier separately can represent a considerable saving but be prepared for a long wait before you assemble a complete set, if you decide to take this approach. The woodwork may not match too well either, though not much of it is visible if the units are stacked.

Television and Exports

The diverse range of television broadcasting standards in use across the world meant that exporting TV sets was rather more difficult than exporting radios, record players and tape recorders. The arrival of colour in the late 1960s changed the marketplace greatly, however, as suddenly there was a huge demand for sets, opening up markets and giving access to large numbers of potential customers who were keen to buy colour sets, despite their very high price.

B&O did not rush to produce a colour set. Their first model, the Beovision 3000 appeared in 1968 and was well-received. A well thought-out, 'no compromise' design, it was capable of outstanding results of a standard that one would expect from a broadcast monitor, rather than a domestic set.

Once the designers had got to grips with the complexities of producing a colour picture, they went on to perfect it. The Beovision 3000 included elaborate circuitry that ensured that the picture was square, linear, of a constant size and correctly coloured. Many early colour sets from other manufacturers could not truly manage any of these things and so the Beovision 3000 really stood out. Perfection came at a cost, however, for the Beovision 3000 was massively complex, even by the standards of the day. It used special transistors in its receiver and colour decoder stages, and

RIGHT: **A side-on view shows the great depth of the cabinet necessary to house the early types of colour picture tube.**

ABOVE: **The Beovision 3000 was the first B&O colour television set.
Availability** ●
Complexity ●●●●

LEFT: **The three coloured circles represented the latest technology but the quality of the cabinetwork could not have been more traditional.**

1964 TO 1969: NEW MARKETS, NEW TECHNOLOGIES   39

The demand for colour meant that luxury monochrome sets like this 24in Beovision 1400SJ, with its rosewood finish and tambour door, would soon disappear from the range. They are amongst the most difficult products of this period for collectors to find today.
Availability ●
Complexity ●●●●

valves for the high-power sections, such as tube drive, scanning, high voltage and audio output. Breaking away from normal television design practice, the line scan circuit and the high-voltage generator were made separate. This allowed the function of each to be optimized, improving the picture shape and stability. The sound section was not neglected (as it so often is in televisions), for the Beovision 3000 was fitted with a comprehensive pre-amplifier complete with treble and bass controls, a powerful valve output stage and a quality loudspeaker system fitted in an enclosure under the screen.

The Beovision 3000 was designed for the 625 line standard only and was fitted with a four-band UHF/VHF tuner that covered most European broadcasting standards. Exporting the sets to Great Britain was another matter, however, as the two most popular programmes were still broadcast only on the old 405 line standard in some areas, obliging British manufacturers to produce elaborate dual-standard sets that could receive both services. In general, imported sets were not a feature of the British television market in the late 1960s; Sony of Japan was the only foreign producer of dual-standard sets. Philips, for example, opted instead to design and manufacture suitable models locally. Even though the Beovision 3000 could only receive 625 line programmes, it was still decided to offer it on the British market. Its outstanding quality produced a surprising number of sales nevertheless, though it was a matter of slight embarrassment that one of the most expensive sets available could only receive one programme. By the end of the 1960s the situation had improved and all three programmes were transmitted on the 625 line colour service, bringing the dual-standard era to a close.

The success of the Beovision 3000 created a market for B&O's monochrome models in Great Britain as well. The Beovision 1400 series, typical of what was available in this period, was a high-quality 24in set that was offered in a range of cabinet styles. A particularly novel feature of this model was that it was amongst the first to use electronic tuning, a feature that would become universal in the following decade.

The Beovision 3000 is for serious collectors only, as it is very large, very heavy and very complicated. The high-voltage circuitry can be lethal in unskilled hands;

even professional engineers feared the sets when they were new. Working examples tend to be in the care of those with particular television skills, who are able to cope with the regular maintenance that all early colour TVs required. The large monochrome models of this period are now very rare and a great deal of luck is required to locate one of the few survivors.

New Models for 1969

As the 1960s drew to a close, B&O found themselves in an enviable position. The plan to save the company by moving upmarket and by becoming internationally recognized had succeeded and the new ranges of models were selling well. 1969 saw some important new models that would continue the success and make B&O the truly global marque that it is today.

Most important of these was the Beomaster 1200. This compact AM/FM receiver took the slide-rule concept to its ultimate conclusion by spreading the tuning dial across much of the top of the unit, complete with a giant slide cursor for station selection. The Beomaster 1200 could be laid flat on a table, propped up using the built-in stand or hung on a wall. Its internal workings included a 15W stereo amplifier, an AM/FM radio with three pre-set FM programmes and an integrated stereo decoder, all constructed using the latest silicon transistors.

For the first time, a matching set of additional components was also offered; these would later include the Beogram 1200 turntable (a development of the Beogram 1800, also introduced in 1969) and the Beocord 1200 open-reel tape recorder. Matching Beovox 1200 loudspeakers were also offered, which had pressed metal grilles and were slim and wider than they were tall; they should not be confused with

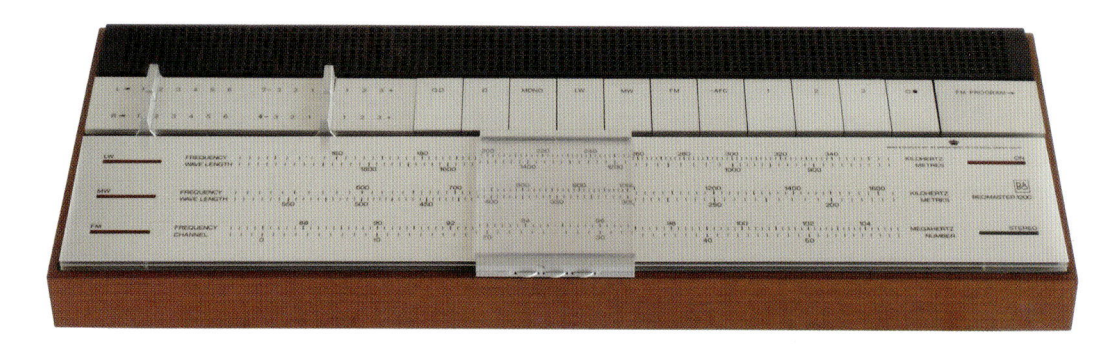

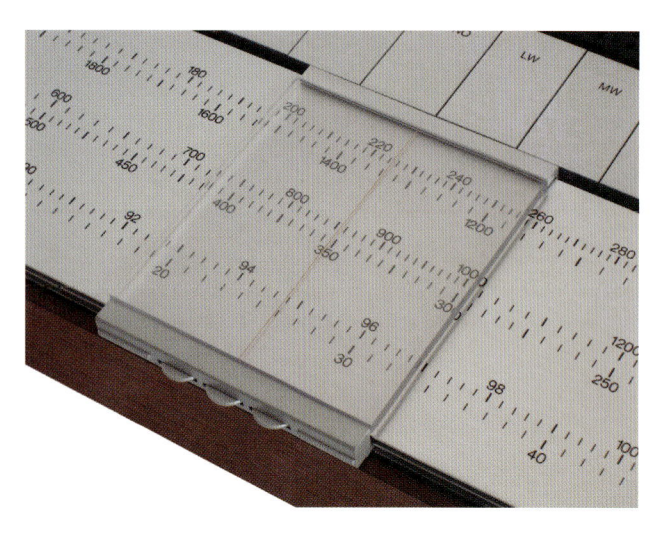

ABOVE: The Beomaster 1200 was considered to be such an original design that it was instantly made part of the permanent collection of the New York Museum of Modern Art (MoMA), along with six other B&O products of the same period.
Availability ●●●●
Complexity ●●

LEFT: This slide cursor was used to set the radio tuning. Note the three thumbwheels for fine adjustment.

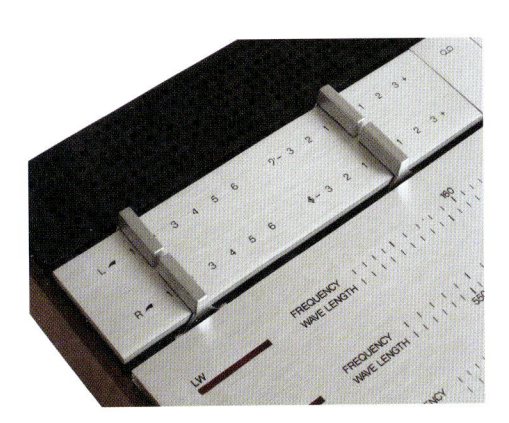

The amplifier controls were designed so that they always looked neat when set for normal operating conditions.

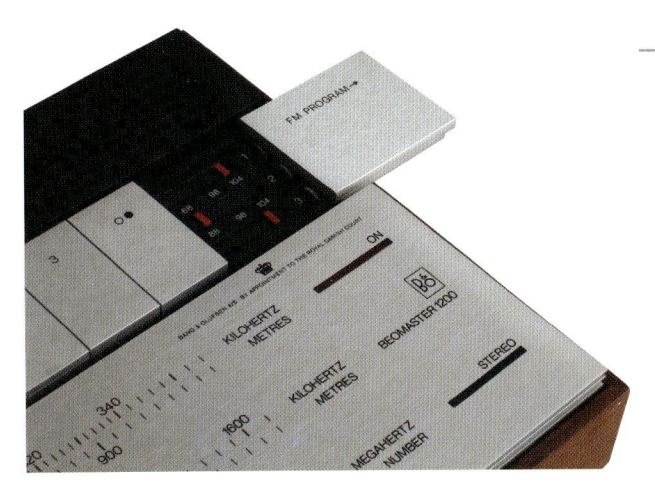

This metal cover, which conceals the pre-set tuning controls, is easily lost and the chances of finding a replacement are thin.

another Beovox 1200 model that was produced a few years earlier.

The 1200 system is an ideal collector's piece and, with the exception of the tape recorder, all the items are fairly easy to find, so it pays to be discerning. The slide cursor and pre-set tuning cover of the Beomaster 1200 are often missing and are difficult to replicate in a satisfactory manner, so ensure these are present and correct. The turntable is unlikely to operate correctly without some kind of mechanical servicing and while the stylus can still be obtained, it is expensive, so budget accordingly.

1969 also saw a new range of tape recorders that replaced the Beocord 1500 and the Beocord 2000 De Luxe. Known as the Beocord 2400 and Beocord 1800, these new models looked very similar but were fitted with completely revised mechanicals and new all-silicon electronics. They were well-respected and won an award for industrial design.

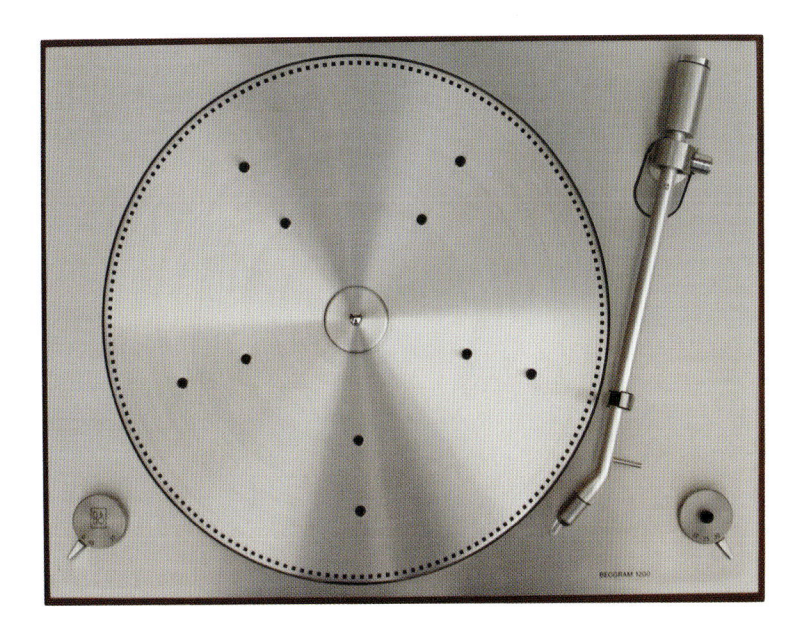

The most beautiful record player in the world? Functionality and quality combined perfectly in the Beogram 1200.
Availability ●●●●
Complexity ●●

1964 TO 1969: NEW MARKETS, NEW TECHNOLOGIES

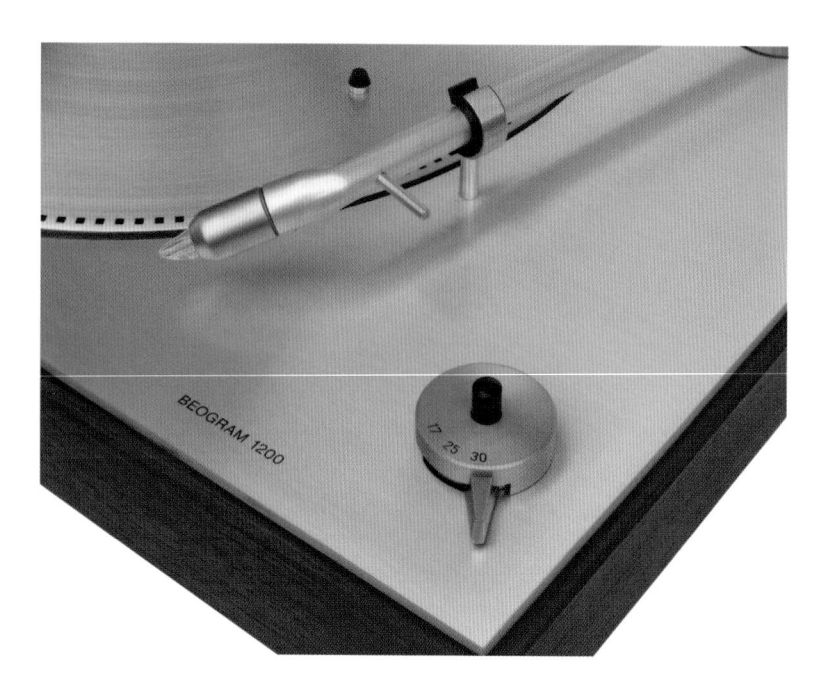

ABOVE: The Beogram 1200 was fully automatic in operation – selecting the record size and pressing the black button were all that was needed to start the mechanism.

TOP LEFT: The Beogram 1200 used a new range of high-compliance SP pickups that had also been restyled to make the arm appear even more elegant.

LEFT: The speed selector included a neutral position to protect the idler wheel during long periods of disuse.

The Beomaster 1600 replaced both the 1400 and the 900. Still of roughly the same dimensions, and available both with and without built-in loudspeakers, this high-quality model featured comprehensive short-wave coverage, complete with a fine tuner.

Most impressive of the new range, however, was the Beomaster 3000. This model was B&O's second entry into 'high fidelity' and was in simple terms a smaller, less complex Beolab 5000/Beomaster 5000 combination fitted into a single cabinet. Although the 30W stereo amplifier was not as powerful as that of the Beolab 5000, the other aspects of the Beomaster 3000's performance were well up to scratch. Important lessons had been learned from the Beolab project and these allowed the Beomaster 3000 to be a more developed, more focused product. An important development was the first use by B&O of integrated circuits in the FM radio circuit. These were sourced from the Radio Corporation of America (RCA) and enabled the designers to make an FM radio as good

RIGHT: **The Beocord 2400 won a coveted Industrie Forme prize at the 1969 Hanover Fair. It would be the last open-reel Beocord aimed at the semi-professional user.
Availability ●●
Complexity ●●●●**

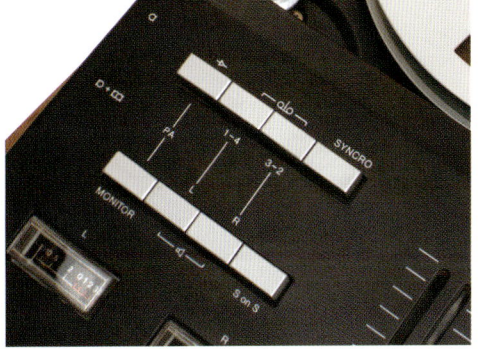

ABOVE: **Separate recording and playback amplifiers, operated by these controls, made for great flexibility and high quality.**

ABOVE: **The carefully laid out and clearly marked control panel gave the Beocord 2400 the feel, as well as the performance, of a proper professional machine.**

LEFT: **The small wheel in this picture is called a slack absorber. It ensures that the tape tension and speed are always perfectly constant.**

1964 TO 1969: NEW MARKETS, NEW TECHNOLOGIES

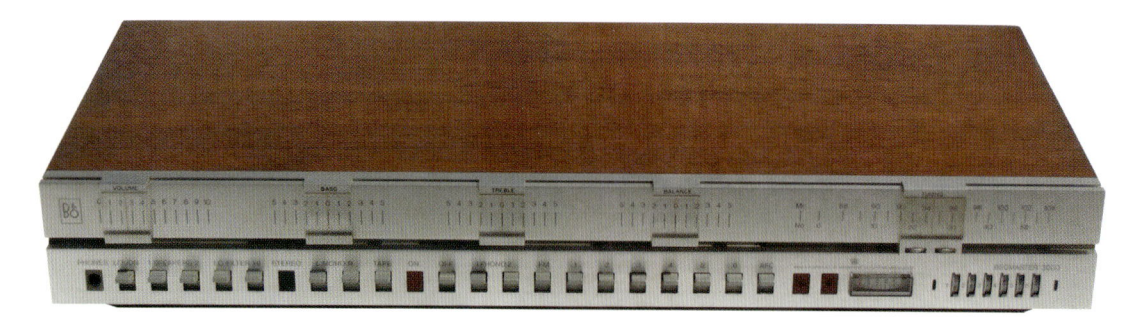

Advancing technology made it possible to produce a compact receiver with true high fidelity performance. The Beomaster 3000 was an instant success and quickly became the standard against which others were judged.
Availability ●●●●●
Complexity ●●

The Colouradio range offered a choice of both models and colours.
Availability ●●●●
Complexity ●

as the Beomaster 5000 on a single printed circuit panel, roughly the size of a compact disc case. As the radio used electronic tuning, it was easy to include six pre-set programmes, a feature that the Beomaster 5000 did not have. The amplifier also used RCA transistors and featured comprehensive facilities, such as noise filters, multiple inputs, a properly implemented tape loop and elaborate stereo/mono switching. The same type of slide-rule controls were used to demonstrate continuity from the Beolab system and these were augmented by nineteen miniature aluminium keys that operated all the switching functions.

The Beomaster 3000 and its derivatives would remain in production in one form or another for ten years, a major achievement in such a fast-moving market. It is an ideal entry into collecting B&O equipment and, at the time of writing, quality examples can be bought quite cheaply, although this may not be the case forever. They combine timeless styling with performance that is well up to current standards and, with no real fears concerning reliability, they can be used every day without worry. A Beomaster 3000 would make an excellent alternative to a Beolab 5000 system, if funds or space are limited.

1964 TO 1969: NEW MARKETS, NEW TECHNOLOGIES   **45**

BELOW: The Beolit 500 was an FM-only model that could be operated on either batteries or mains power.

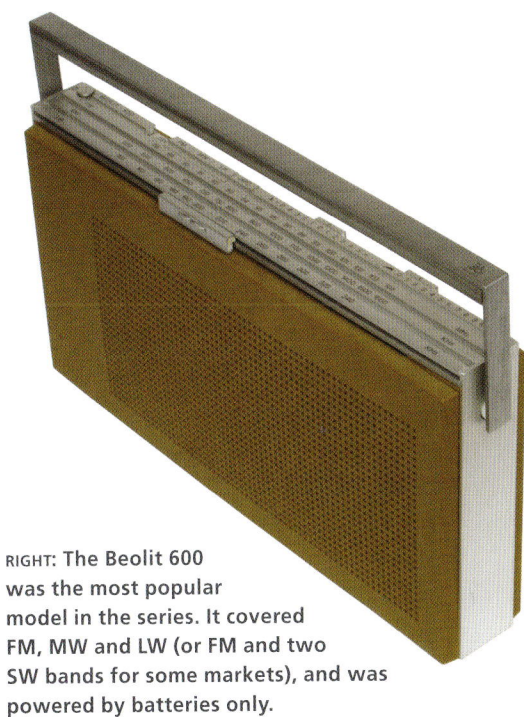

RIGHT: The purple finish of this Beolit 400 was not the most popular choice when new but is in great demand by collectors now.

RIGHT: The Beolit 600 was the most popular model in the series. It covered FM, MW and LW (or FM and two SW bands for some markets), and was powered by batteries only.

ABOVE: The Beolit 700 was the top of the range model and added battery or mains operation to the Beolit 600 specification.

46   1964 TO 1969: NEW MARKETS, NEW TECHNOLOGIES

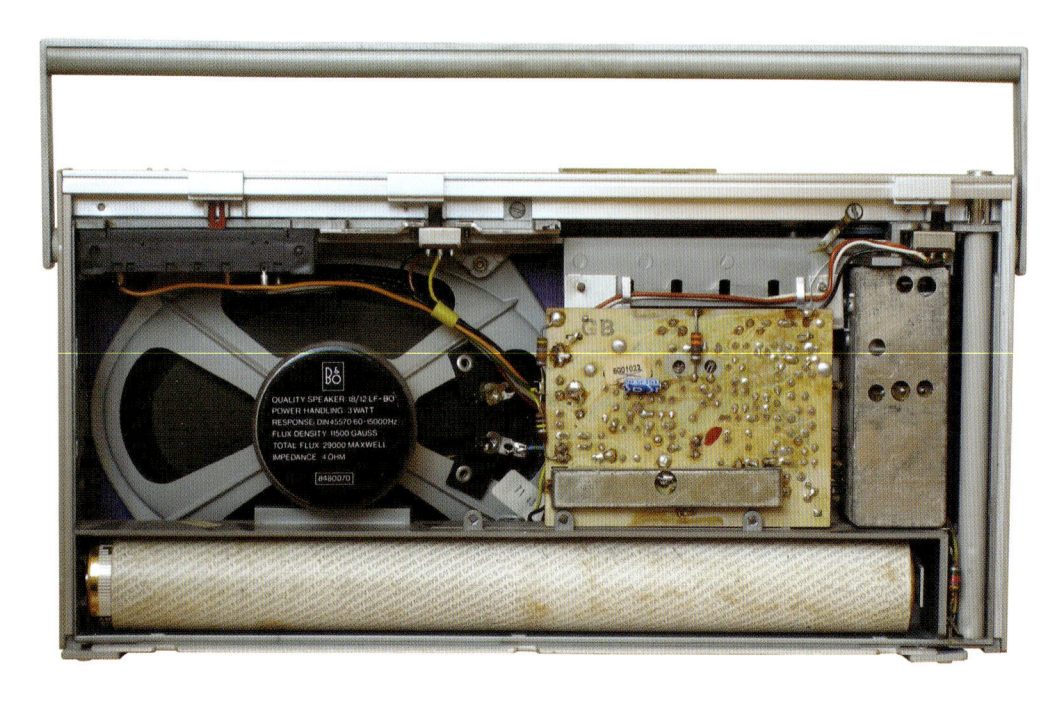

This view inside the Beolit 400 shows the unusually large loudspeaker. This model was the simplest in the range and covered the FM band only.

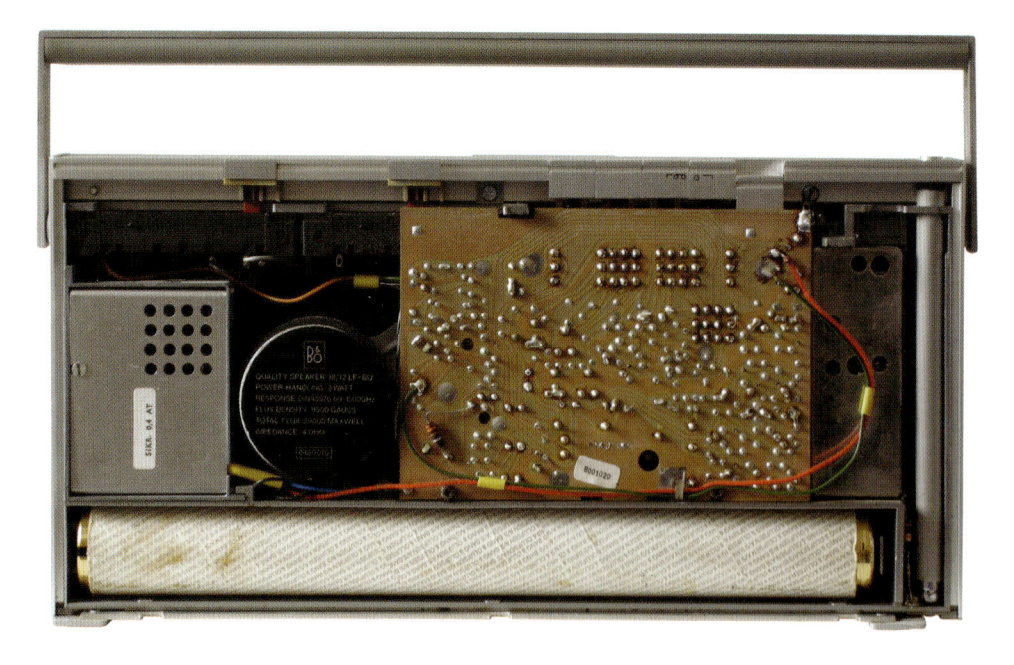

In contrast to the Beolit 400, the Beolit 700 contained a more complex printed circuit and a mains unit (the silver box on the extreme left).

1964 TO 1969: NEW MARKETS, NEW TECHNOLOGIES   **47**

Also shown in 1969, in preparation for sale in 1970, was a new type of Beolit portable called the Colouradio. This attractive set, much loved by collectors now, replaced all the previous Beolit ranges with one simple series of models that offered either AM/FM or FM-only reception, and battery-only or mains and battery operation, depending on which of the four models was chosen. The name came from the clip-on plastic side panels that were available in a choice of red, black, white, violet or 'curry'. These could be changed by the owner without the use of tools, just like the similar schemes offered by mobile telephone manufacturers about three decades later.

These high-quality sets remain very much in demand and can be quite expensive but don't rush into a purchase. Make sure that the antenna is present

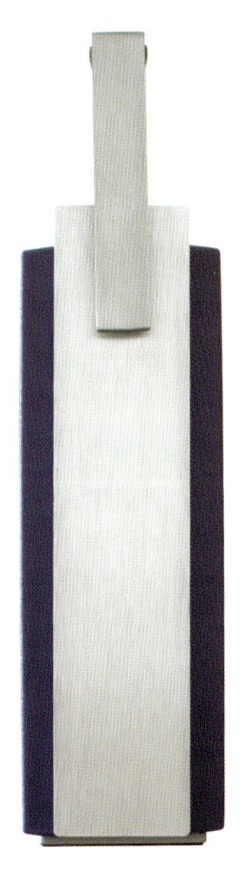

LEFT: Viewed from the side, the Colouradio was very slim.

RIGHT: The same view of one of the mains/battery models shows the location of the power socket.

BELOW: Two different tuning cursors were used on the Colouradio models, the clear plastic type (left) for the Beolit 400 and 500 and the magnetic ball bearing type (right) on the 600 and 700.

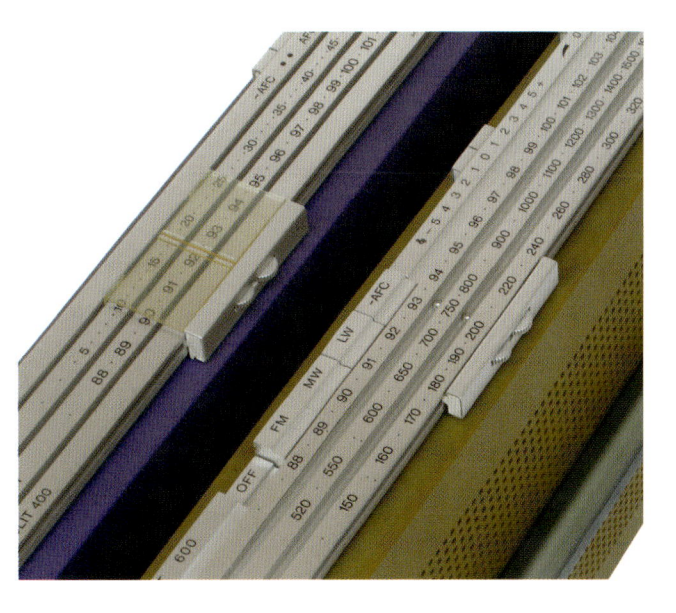

and straight, all the sliding control knobs are still there (complete with their metal trims) and that the battery holder is not excessively corroded. The loudspeaker is a weak point and is not easily substituted, so if the sound is not clear and rich, proceed with care. Some of the bright cabinet colours fade or discolour with age (red and white are the worst in this respect) but careful repainting can restore the finish. Models with built-in mains adaptors (the Beolit 500 and Beolit 700) are the most practical for everyday use, but ensure a decent mains lead is included, as not all the alternatives that are currently available fit properly.

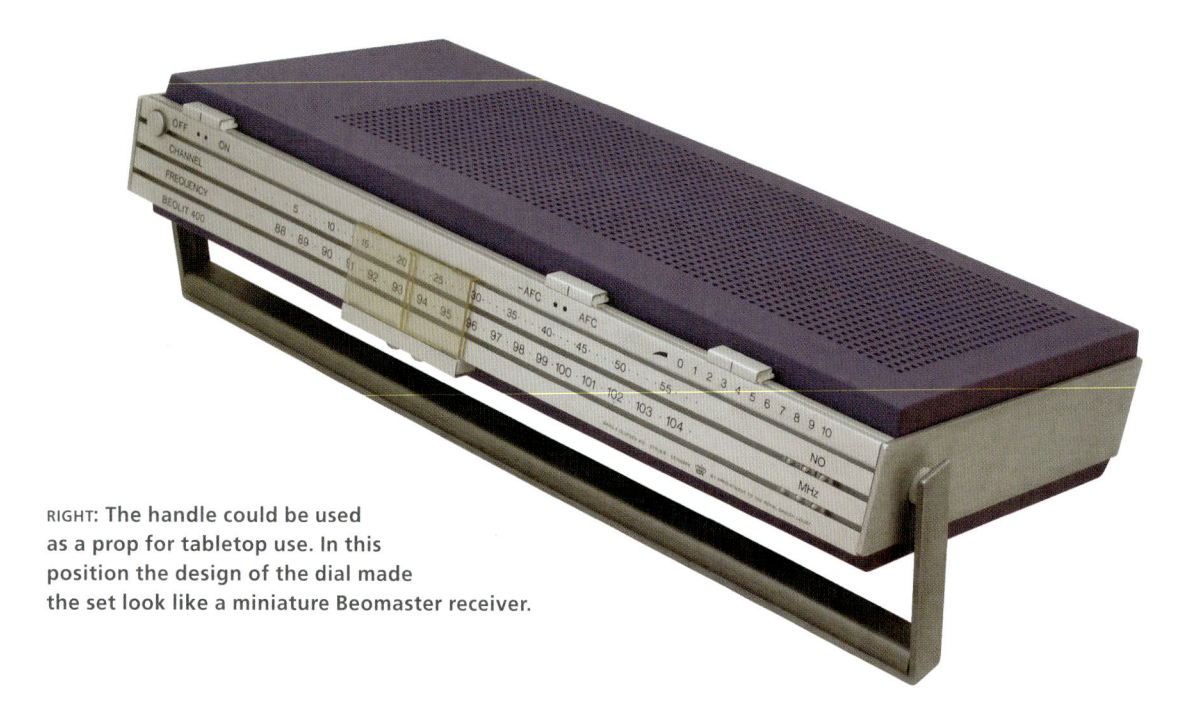

RIGHT: The handle could be used as a prop for tabletop use. In this position the design of the dial made the set look like a miniature Beomaster receiver.

LEFT: The ball bearings are moved by a magnet beneath the fascia and held in place by a strip of clear plastic.

CHAPTER THREE

1970 TO 1979: A DECADE OF ACHIEVEMENT

Perfecting the Beomaster

The Beomaster receiver was a key product for B&O, as every system, no matter how basic, had to have one. A system could work without a turntable or a tape recorder but would always need an amplifier of some kind.

The first major event of the 1970s for the Beomaster line occurred in 1971 when the Beomaster 3000 became the Beomaster 3000-2. The small name change primarily indicated more power, as the amplifier could now produce 40W instead of the previous 30W. The Beomaster 3000-2 was so good that the Beolab 5000 system was quietly dropped from the range during this period. This downgrading in the power of B&O's top system did not last for very long and 1973 saw the introduction of the Beomaster 4000, essentially a 60W version of the Beomaster 3000-2. Distinguished by its black fascia and black metal grille at the rear, the Beomaster 4000 was not as popular as could have been expected – evidently 40W was enough.

The final version of the Beomaster 3000 theme arrived in 1977. The Beomaster 4400, apart from its angled front panel, looked very similar to the original model but inside everything was new. The amplifier produced 75W, well over twice as much as the original Beomaster 3000. Not only is the Beomaster 4400 one of the most powerful Beomaster models, it is also the best sounding. Its outstanding performance demonstrates how little (if any) progress has been made in amplifier design in the last thirty years.

The smaller models had also been under development. The most important of these was the Beomaster 901, which appeared in 1973. Styled by Jacob Jensen and looking a bit like a shrunken Beomaster 3000, the Beomaster 901 set the pattern for the smaller models for years to come.

Elegant and compact, the Beomaster 901 was rated at 20W and covered the LW, MW and FM stereo wavebands. Later, an FM-only version would appear, the Beomaster 1100. To make up for the lost wavebands, the Beomaster 1100 offered four pre-tuned pro-

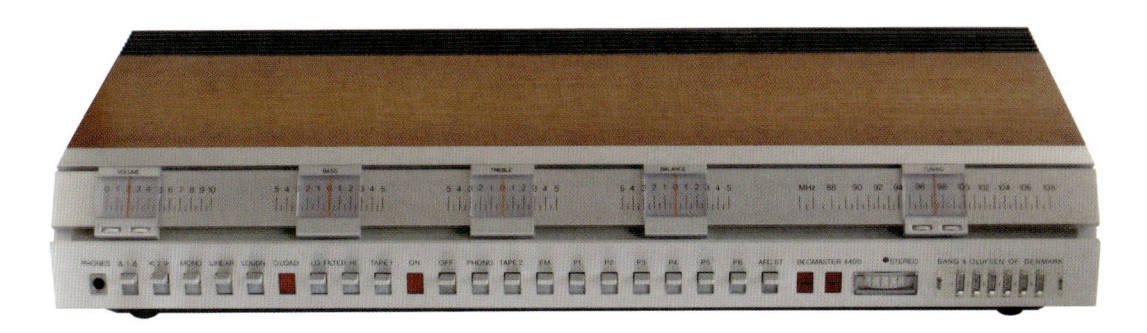

This is quality. The powerful Beomaster 4400 is the audiophile's choice in the Beomaster range.
Availability ●●●
Complexity ●●

1970 TO 1979: A DECADE OF ACHIEVEMENT

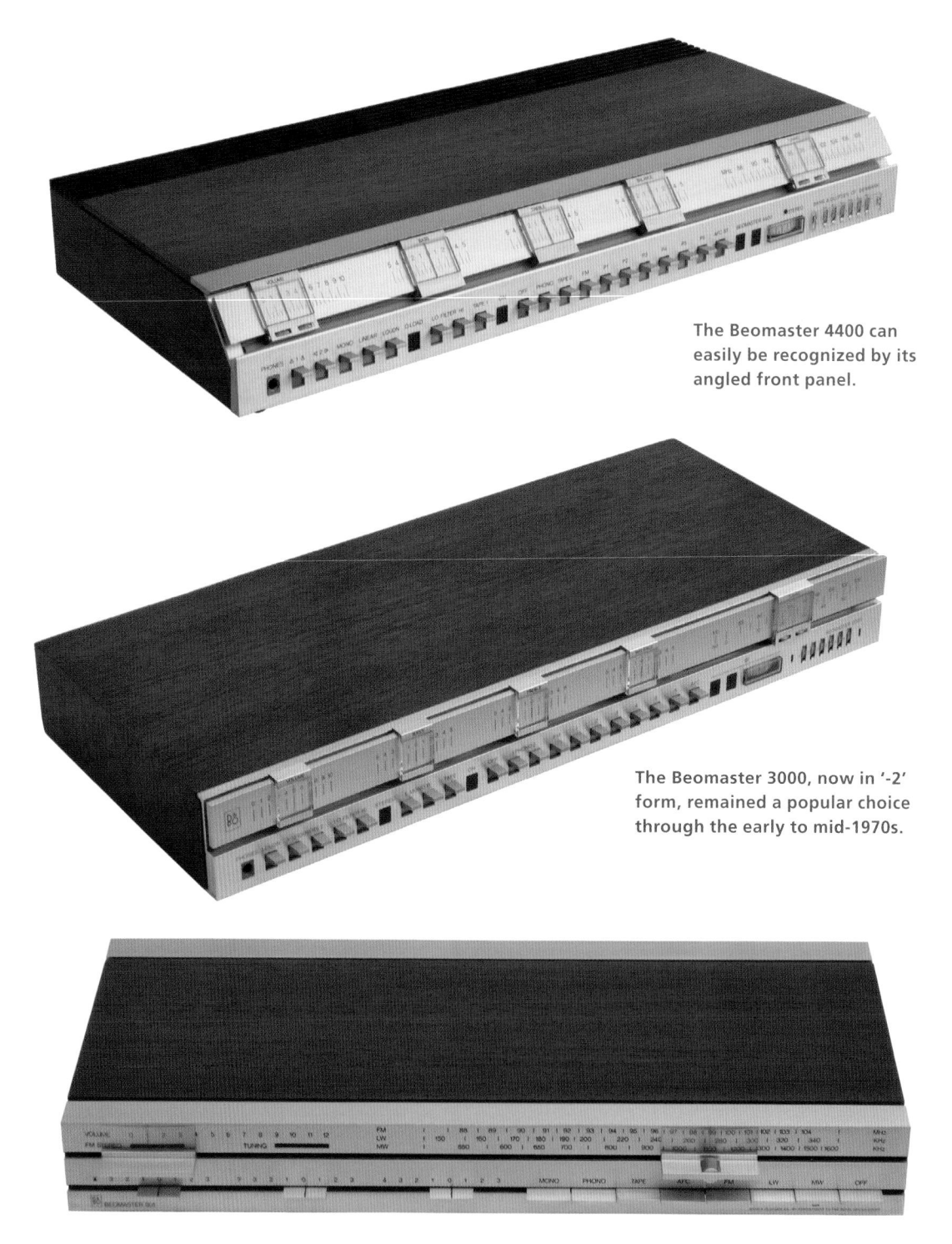

The Beomaster 4400 can easily be recognized by its angled front panel.

The Beomaster 3000, now in '-2' form, remained a popular choice through the early to mid-1970s.

The Beomaster 901 epitomized the new B&O look that would endure throughout the decade.
Availability ●●●●
Complexity ●

1970 TO 1979: A DECADE OF ACHIEVEMENT   **51**

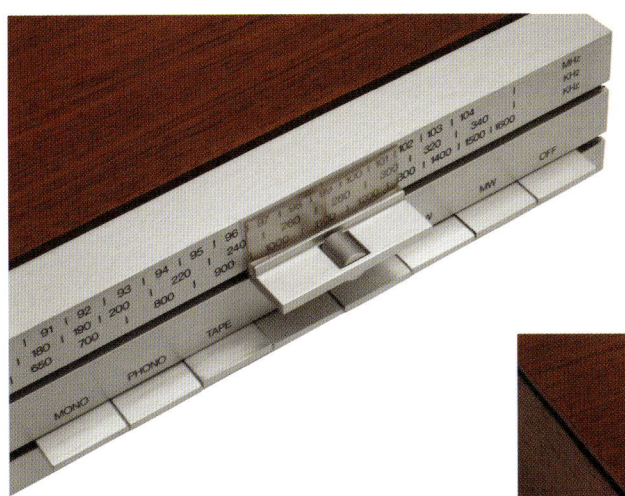

LEFT: The Beomaster 901 tuning control was operated either by sliding the cursor or by making fine adjustments with the small roller in the centre.

BELOW: The volume control was styled to match the tuning cursor.

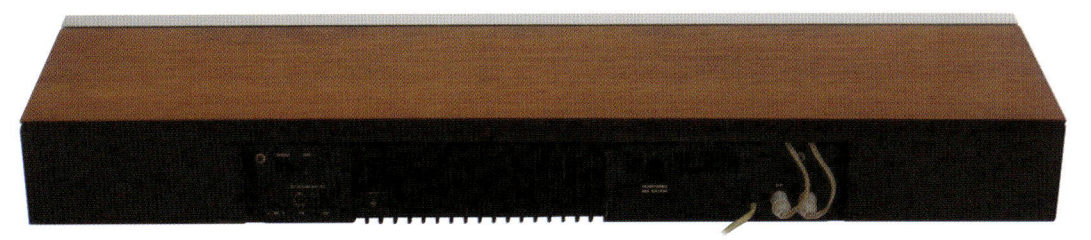

ABOVE: This rear view of the Beomaster 800 shows how a standard Beomaster 901 chassis was fitted into the larger cabinet.

ABOVE: The Beomaster 901 also formed the basis of the last B&O table radio, the Beomaster 800.
Availability ●●
Complexity ●

1970 TO 1979: A DECADE OF ACHIEVEMENT

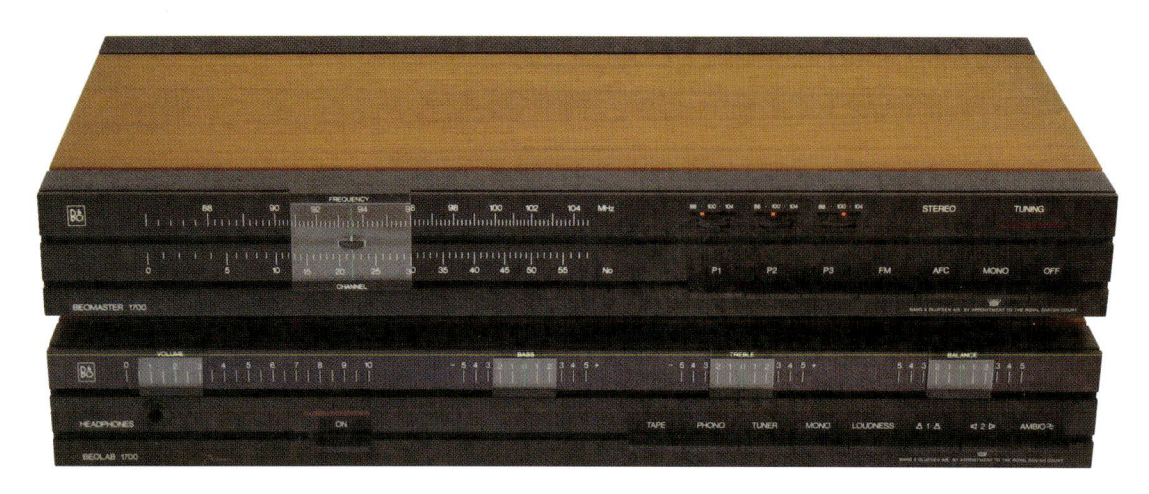

ABOVE: **The Beolab 1700 system of 1973.**
Availability ●●●
Complexity ●

LEFT: These controls look like those of the Beolab 5000 but they are not so well engineered or as pleasant to use.

RIGHT: The first B&O stack system was not a great success and so the company returned to producing its trademark long and low designs.

grammes, slightly more power and a headphone socket sensibly mounted on the front panel. The Beomaster 1100 was finished in black, as had become the pattern for FM-only models.

Before the Beomaster 1100 appeared, however, B&O offered something very similar except that it was split into two boxes. The Beolab 1700 system of 1973 attempted to recall the power and prestige of the original Beolab 5000 system and comprised two very similar looking units: the Beolab 1700 amplifier and the Beomaster 1700 tuner; although direct comparison would reveal that they were much smaller.

Rated at 20W, the Beolab 1700 system was nowhere near as powerful as the original 5000 model and, as the Beomaster 901 offered similar performance in half the space and at similar cost, it is difficult to see why B&O bothered. The system could be stacked neatly with a Beocord 1700 cassette deck on the top – the first occurrence of a stack system in the range. This format was not something that the B&O designers approved of and it would be over ten years until the next stack system appeared.

The Beomaster 1200 got the FM-only black treatment in the early 1970s and became the Beomaster 1001. This formed the centrepiece of a basic budget system that also comprised B&O's last manual turntable, the Beogram 1001 (basically a slightly re-styled Beogram 1000) and a pair of Beovox 1001 loudspeakers, which were a simple model similar to the preceding Beovox 1200 but, of course, finished in black. These loudspeakers still retained obsolete cone-type tweeters and were capable of only moderate performance.

One new feature that the Beomaster 1001 did introduce was ambiophonic sound. This system used an additional pair of small loudspeakers placed at the rear of the room to provide a basic 'surround sound' effect. The other Beomaster models that included this facility were the 4000, 4400, 4401 and 6000 4channel, as

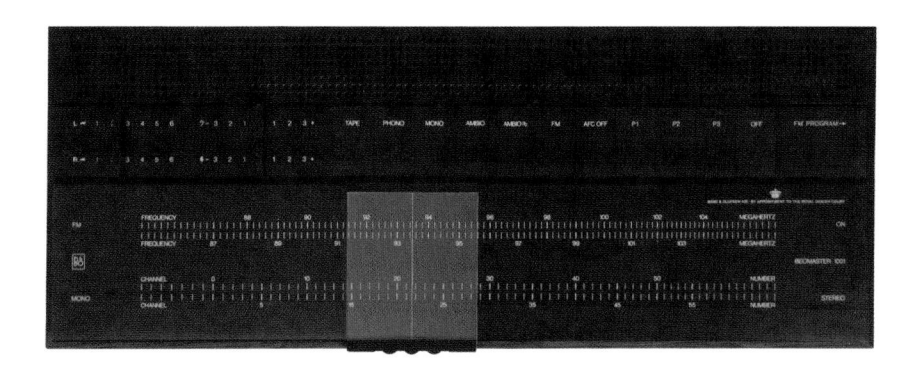

The Beomaster 1001 can easily be distinguished from the better known 1200 model by its all-black finish.
Availability ●●
Complexity ●●

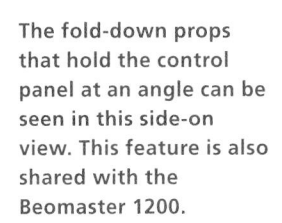

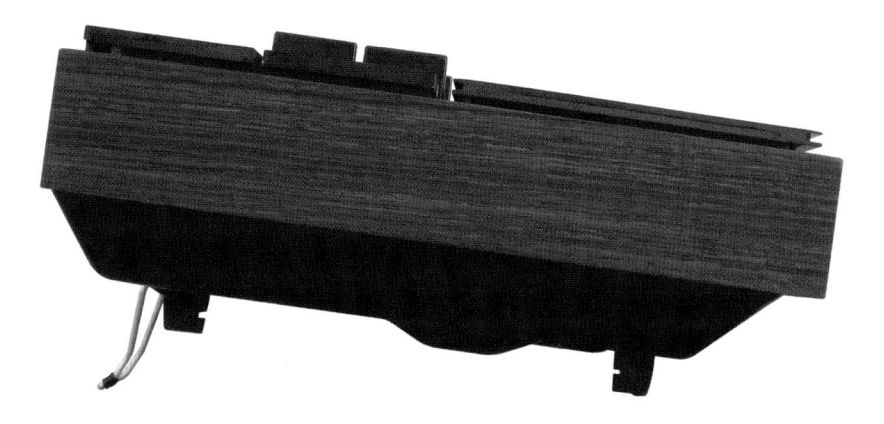

The fold-down props that hold the control panel at an angle can be seen in this side-on view. This feature is also shared with the Beomaster 1200.

well as the Beolab 1700 amplifier. For those without, a small adaptor unit was made available.

A big leap forward in technical terms came with two new models in 1974. First of these was the Beomaster 2000, a sleek, slim model with all its controls on its top surface. The Beomaster 2000 included an important innovation. The power amplifiers it used were of a new design that was directly coupled to the loudspeakers, dispensing with the capacitors and transformers of previous designs. This greatly improved the sound quality that could be achieved and soon became a feature of the great majority of models that followed.

The other new model was the Beomaster 6000 4channel and was B&O's entry into quadraphonic sound, a brief craze of the mid-1970s. Not content with producing a four-channel amplifier, B&O also gave the Beomaster 6000 4channel touch controls for

ABOVE: **The Beomaster 2000 of 1974 introduced new amplifier technology that would soon become standard across the range.**
Availability ●●●
Complexity ●●

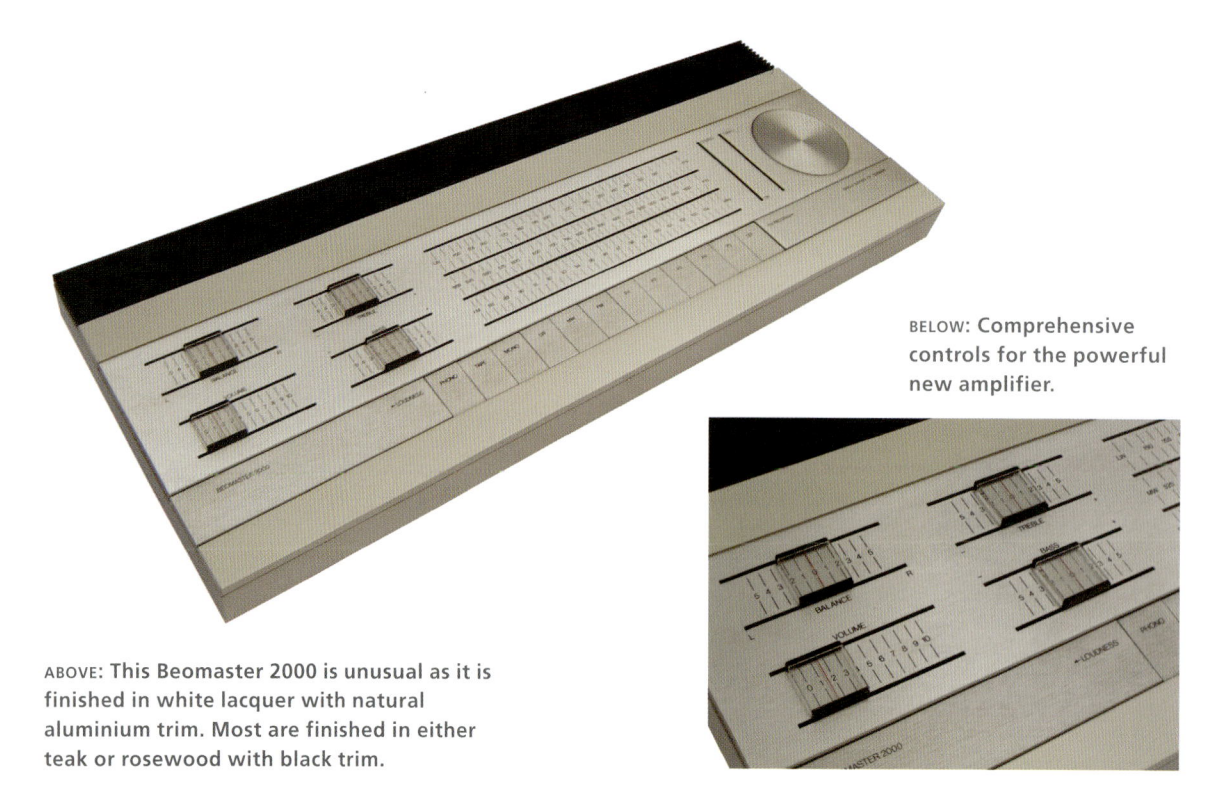

BELOW: **Comprehensive controls for the powerful new amplifier.**

ABOVE: **This Beomaster 2000 is unusual as it is finished in white lacquer with natural aluminium trim. Most are finished in either teak or rosewood with black trim.**

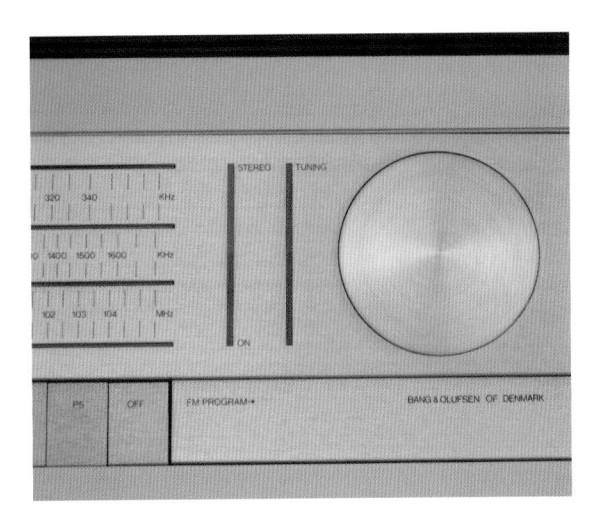

1970 TO 1979: A DECADE OF ACHIEVEMENT   55

The big wheel on the right operated the tuning; it had a free-spinning flywheel action.

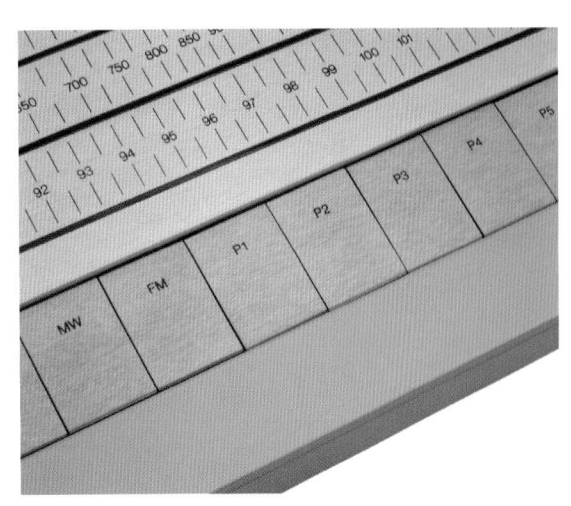

Source and programme selection was by super-smooth metal pushbuttons.

all its major functions. The source-selection buttons worked using clever solid-state signal routing circuits but there was no easy way to make them operate the sound functions such as volume, treble, bass and balance. Amazingly, the solution that was settled on was to use conventional rotary controls but to bury them deep inside the machine. The controls would be operated by an electric motor driving through magnetically operated clutches. As well as driving the controls, the motor also drove illuminated pointer indicators that showed to which position each control was set. The large gloss black section on the top of the cabinet was used as a display, with the various functions illuminating as they were brought into action.

The Beomaster 6000 4channel was a natural candidate for remote control and indeed one was available as an optional extra. The handset was similar to that supplied with the Beovision 6000 colour TV and used ultrasonic sound as the transmission medium.

The Beomaster 6000 4channel included a built-in decoder for 'SQ' format quadraphonic records and could also be used with a quadraphonic tape recorder (though B&O never produced one). B&O's preferred medium for quadraphonic sound, however, was the

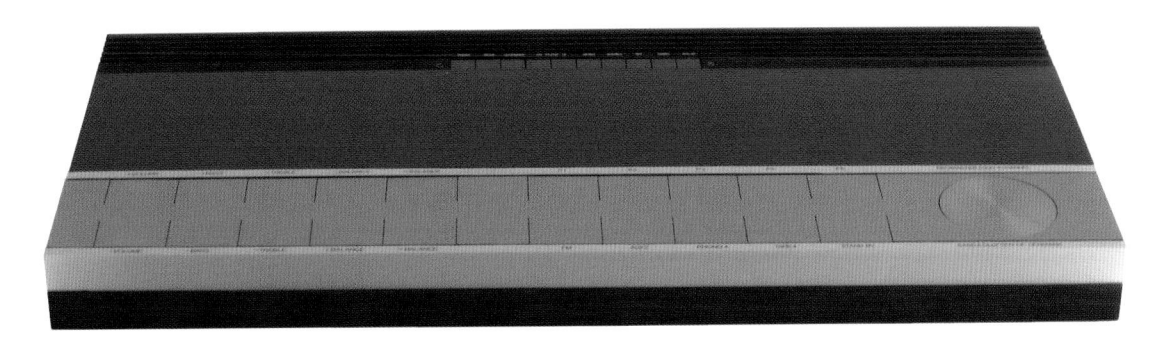

The quadraphonic Beomaster 6000 4channel dispensed with knobs and buttons in favour of touch controls, making it one of the most distinctive hi-fi designs of the 1970s.
Availability ●●
Complexity ●●●●●

A detailed view of the controls of the Beomaster 6000 4channel. Slits in the metal allowed each key to move independently.

Not all of the controls of the Beomaster 6000 4channel were electronic; the tuning, for example, was adjusted with this freely spinning wheel.

LEFT: The Beomaster 6000 Commander had a keypad that matched the design of the controls of the Beomaster. The ultrasonic emitter can be seen clearly in this head-on view.

BELOW: The Beomaster 6000 commander was a substantial unit as this view of it with a more recent Beolink 1000 remote control shows.

CD4-encoded LP, as pioneered by JVC. This required a special pickup and a complex decoder, both of which were included in the matching Beogram 6000 quadraphonic record player.

Quadraphonic sound did not take off in the way that B&O, and the rest of the industry, hoped that it would. B&O produced one more quadraphonic system, the 3400. This was based around the Beomaster 3400, which was a version of the Beomaster 2000 adapted for four-channel operation (something that came at the expense of the AM radio bands), along with the Beogram 3400, a modified Beogram 1900 fitted with a CD4 decoder. Although simpler and cheaper than the 6000 system, the 3400 was more developed and capable of better performance. This did not make it any more of a success, however, and very few were sold.

1976 saw B&O launch another step-ahead signature design, the Beomaster 1900. The derivatives of this model would remain in production until the mid-1990s; the Beomaster 1900 itself sold so strongly that the launch of the remote-controlled version, the Beomaster 2400, had to be delayed until the following model year.

The Beomaster 1900 was slim and sleek in a way that no previous stereo receiver had been before. Two design principles were put to work to achieve this effect. Firstly, the main controls were made touch-sensitive, removing the need for mechanical switches, sliders, knobs or buttons. Instead, the operator merely had to lightly touch the printed legend indicating the required function. Secondly, the controls were divided up into those that would be used on a daily basis and those that were only required occasionally. These two sets of controls were placed in areas known as the primary and secondary operating panels. The primary operating panel was the touch-sensitive strip on the front of the set; the secondary operating panel was

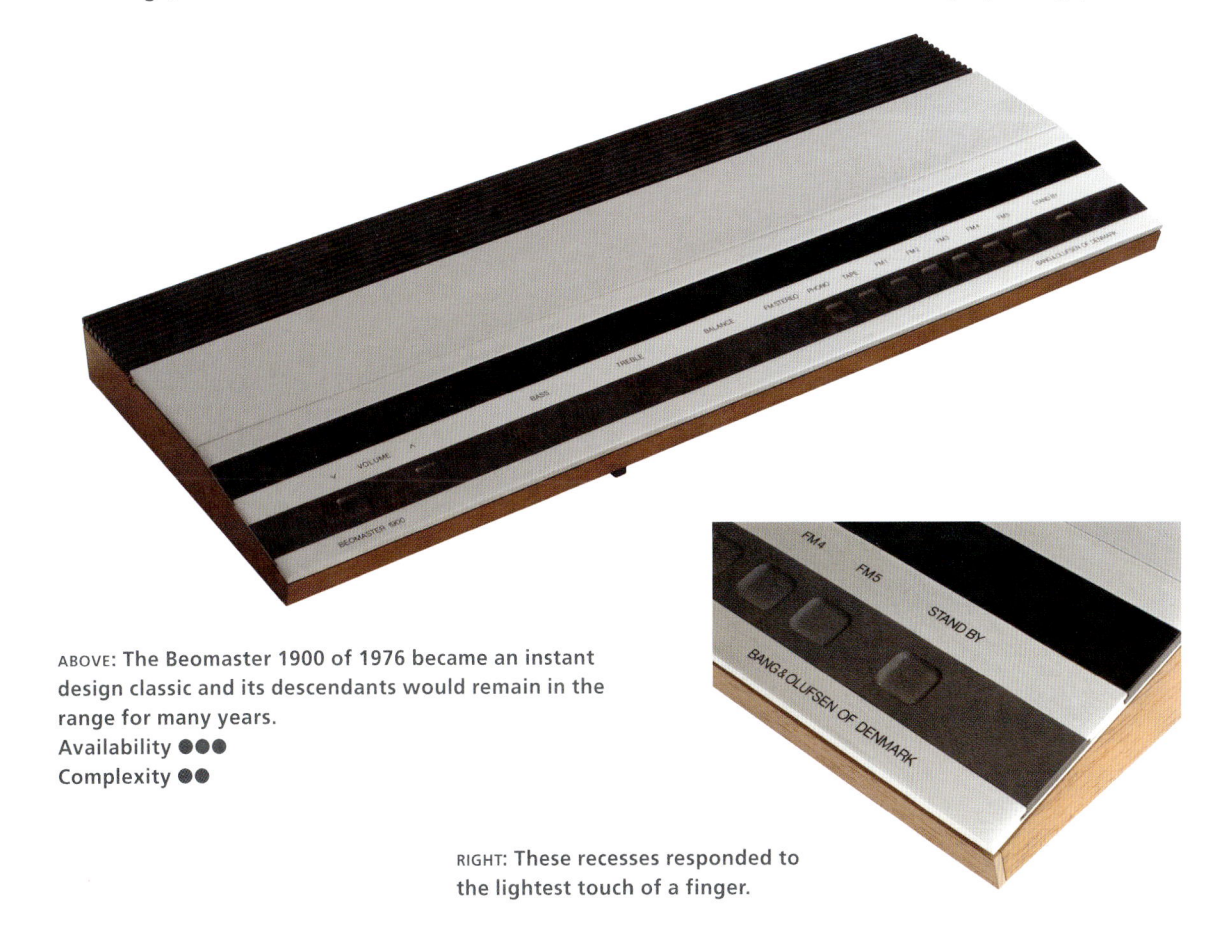

ABOVE: **The Beomaster 1900 of 1976 became an instant design classic and its descendants would remain in the range for many years.**
Availability ●●●
Complexity ●●

RIGHT: **These recesses responded to the lightest touch of a finger.**

1970 TO 1979: A DECADE OF ACHIEVEMENT

With the lid open, the secondary operating panel can be seen.

The Beocord 1900 was styled to match the Beomaster 1900.
Availability ●●●
Complexity ●●

concealed beneath a lift-up aluminium flap on the top.

Although hailed by B&O as revolutionary, both these design principles had already been successfully applied to television sets for years. The Beomaster 1900 was a striking piece of work all the same, especially when wall-mounted on a special bracket.

To begin with, the optional sources for the Beomaster 1900, the Beogram 1900 and the Beocord 1100, were boxy designs that did not complement

LEFT: The styling match was not perfect; the bulky cassette mechanism of the Beocord made it quite a bit taller than the Beomaster.

1970 TO 1979: A DECADE OF ACHIEVEMENT   **59**

the shape of the Beomaster at all. The situation was soon partly remedied with the launch of the Beocord 1900 cassette deck, whose lines and colours matched the Beomaster, even though the whole cabinet was slightly taller and the large mechanical keys were at odds with the smooth control surfaces of the receiver. The Beogram situation would take longer to resolve but eventually the Beogram 2200 appeared, which offered an acceptable match.

The remote-controlled Beomaster 2400 appeared the following year. Similar to the 1900 but with a slightly re-styled primary operating panel, this model included a remote control unit similar to the Feel Commander supplied with the better Beovision TV sets of the period. The remote-control receiver in the Beomaster 2400 was of a similar design to that used in the televisions, although only eight (instead of sixteen) functions were available. The later Beomaster 2400-2 was similar to the original model but gained extra connections in the record player socket that allowed the Beomaster to turn the record player on and off by remote control, provided that either a Beogram 4004 or 2400 was used. This new level of integration gave a small taste of what was to come, a new generation of audio products that could send control instructions, as well as sound, between themselves.

ABOVE: **The Beomaster 2400 brought ultrasonic remote control to the range of smaller Beomaster models.**
Availability ●●●
Complexity ●●

RIGHT: **The simple and neat Beomaster Control Module remote control unit was similar in design and construction to those provided with the Beovision TV sets of the day.**

60   1970 TO 1979: A DECADE OF ACHIEVEMENT

The difference in the styling of the Beomaster 1900 (left) and 2400 (right) can clearly be seen in this picture.

The Beomaster 2200 looked conventional from the outside but inside it employed an unusual modular chassis.

1970 TO 1979: A DECADE OF ACHIEVEMENT   61

B&O ended the 1970s with one final range of Beomaster models. Although visually nothing appeared to connect the Beomaster 1500, 2200 and the previously mentioned 4400, they were in fact based on a modular circuitry concept, where essentially the same electronic building blocks could be configured to provide 25, 40 and 75W, respectively. All offered class-leading performance in their respective price brackets and remain some of the best performing Beomasters ever offered. The inside of the Beomaster 2200 was

The modular chassis of the Beomaster 2200.

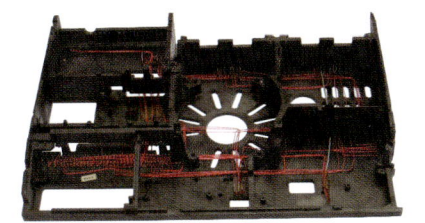

Supporting frame added.

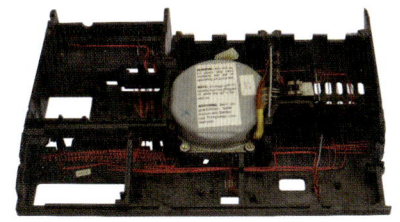

Mains transformer and fuses installed.

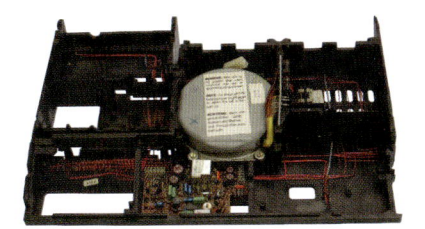

FM stereo decoder fitted.

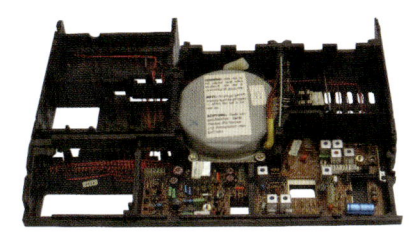

AM radio fitted.

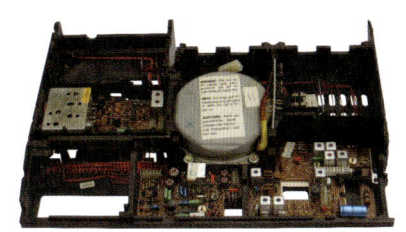

FM radio added.

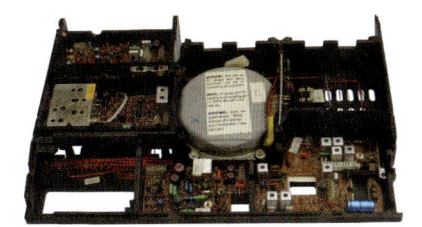

Socket panel and pre-amplifier in place.

Pre-set tuning bank installed.

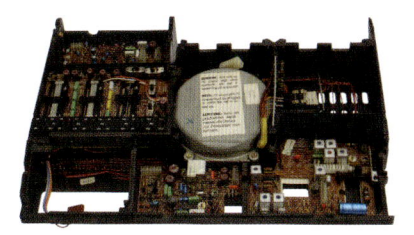

Audio controls added.

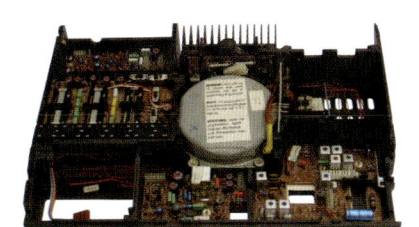

Left channel power amplifier installed.

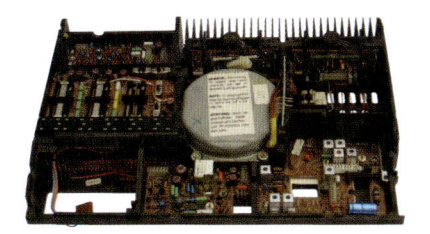

Right channel power amplifier installed.

Power supply regulator fitted.

constructed in a novel manner. The chassis consisted of a resin 'pegboard' onto which numerous small circuit cards were plugged; such practice echoed television techniques and in theory made for easier servicing, although in truth audio units typically needed fewer repair visits than televisions so, in many cases, this careful design was wasted. This method of construction was not used again in a B&O audio product.

Two less frequently seen models in this series are the Beomaster 1400 and the 4401. The Beomaster 1400 was an FM-only version of the 1500 and was, of course, finished in black instead of natural aluminium. The Beomaster 4401 was also finished in black but in all other respects was identical to the 4400.

Beogram 4000

If the Beomaster 900 had demonstrated B&O's intentions to produce up-market products with international appeal, then the Beogram 4000 proved beyond any doubt that they had succeeded in that aim. When first announced in 1972, the Beogram 4000 stunned the industry – nothing like it had been seen before and even today its impact remains undiminished. Beautiful to look at, delightful to use and capable of the most perfect reproduction, no other turntable, regardless of price, can match it.

In its infancy, the Beogram 4000 was known simply as the 'high-fidelity gramophone'. The aim of the project was to produce a record player of the same

Every collector would like to own a well-preserved Beogram 4000 like this one. The perfect styling is matched by flawless build quality.
Availability ●●●●
Complexity ●●●

standard as the rest of B&O's high-fidelity line, epitomized by the Beolab 5000 system of 1967. Lacking the necessary mechanical expertise for such an exacting and ambitious project, B&O hired Gustav Zeuthen, a mechanical engineer with a background in fields as diverse as aircraft and office machinery. Zeuthen already owned a Beogram 1000 and so was well aware of the strengths and weaknesses of existing B&O gramophone technique. Also on the team was Subir Pramanik, who had mastered the mathematics behind ensuring stability in turntable mechanisms. He would also design the radical new pickup that the Beogram 4000 would eventually use.

Meanwhile, a separate high-fidelity gramophone project had been started by Jacob Jensen, the influential designer who had already been responsible for previous successes such as the Beolab 5000 and the Beomaster 3000. Jensen's machine was of traditional construction and used a very long pickup arm in order to reduce tracking errors. The Zeuthen/Pramanik machine was completely new and used a short, stumpy arm that moved with a parallel action. This small arm had a number of technical advantages. It could be made light, stiff and largely free of internal resonances – all important attributes for a high-performance record-playing system. Jensen was not impressed and is said to have commented that 'such a small willy doesn't communicate potency'. The conflict was eventually settled by Jens Bang, who realized that the advanced technology of the parallel tracking machine far better suited B&O's new agenda. He was right. Had he gone with Jensen's design then what resulted would have probably been just another high-end turntable of the early 1970s. Instead, the Beogram 4000 became an instant design classic that was (and remains) loved, recognized and respected throughout the world.

The Beogram 4000 that emerged positively overflowed with inventiveness, innovation and new technology. Every aspect of the functioning of a gramophone had been analysed and rationalized to yield an optimum solution. The most striking feature was, of course, the arm. Parallel and tangential tracking arms had been attempted before but what made the B&O arm novel was that, instead of being dragged along by the grooves of the record, the arm was controlled by an electronic servo. Optical sensors in the arm base detected any angular error that resulted as the stylus followed the groove, and through a complex electronic circuit, a precision motor turned a lead screw that moved the arm. The raising and lowering of the arm was also electronically controlled by a solenoid, whose abrupt action was damped by an air cylinder.

The arm was tipped with Subir Pramanik's new pickup. Following the same logic that had lead to the design of the arm, it was realized that miniaturizing the parts of the pickup could have the same beneficial effects. In principle, the SP 15 pickup used in the Beogram 4000 was a miniaturized version of the latest SP 12 model. Making the parts this small meant that it was not really practical to make the stylus replaceable by the user, so the new pickup was sold as a sealed 'integrated' unit. Once the stylus was worn out, the whole pickup had to be replaced, although B&O did offer a concession of a half-price replacement if the old pickup was returned. While this made running the Beogram 4000 more expensive than previous models, it did ensure absolute accuracy and protected the owner from being sold inferior pattern parts. The pickup pressed home into the arm using a special socket; this simultaneously made all the necessary electrical connections and ensured correct mechanical alignment. The SP 15 was after a short time renamed MMC 4000, MMC standing for 'moving micro cross'.

The MMC 4000 integrated pickup, as used in the Beogram 4000.

64 1970 TO 1979: A DECADE OF ACHIEVEMENT

Replacement MMC 4000 pickups came attractively packaged complete with simple tools for adjustment and cleaning.

A close-up of the control panel of the Beogram 4000. The orange strobe lamp is visible through the small window at the top left of the panel.

The turntable motor of the Beogram 4000 was also a new design. Rather than relying on the consistent frequency of the mains supply to regulate the speed, the Beogram 4000 contained its own AC generator and power amplifier to drive the motor. As well as making the motor independent of the different AC frequencies and voltages used in the various export markets, this technique resulted in smooth and consistent running, and also allowed the turntable speed to be selected electronically without the use of moving parts. Illuminated controls were fitted on the operating panel, so that the user could 'fine tune' the speed, perhaps to accompany a musical instrument that was slightly out of tune. A stroboscopic ring and neon lamp were fitted so that the correct speed could be returned to, but to avoid cluttering the clean lines, the image was printed on the underside of the platter and viewed through a small window on the operating panel, through a series of mirrors.

The motor drove the heavy platter through a wide, flat belt. The platter was so heavy that it required a special oversize bearing to support it. This in turn required a substantial die-cast sub-chassis and a solid plinth to hold it in. Such was the strength of the complete machine that some dealers were moved to demonstrate it by placing it on the floor and standing on it!

Much work had gone into how the sub-chassis was to be supported in the plinth. The final answer came in the form of long strips of spring steel from which the arm and platter assemblies were hung on steel wires. The isolation that this arrangement offered was so effective that B&O coined a new phrase for it: dance-proof.

Since all the important functions of the Beogram 4000 were electronically controlled, it made sense to tie them all together and make operation completely automatic. Automatic record players and record changers had already existed for many years but they were frowned upon by serious listeners, as the gears, levers and mechanical linkages that moved the arm were considered too crude in action to be used with delicate arms and pickups. In the case of the Beogram 4000, these mechanical parts were dispensed with in favour of a digital electronic control unit. Constructed from the latest integrated circuits, this section of the machine could justifiably be referred to by the standards of the day as a small computer. All one had to do to play a record was to place it on the turntable and press 'on'. From that point forward, the Beogram 4000 would automatically select the right record size and speed, play the record and then switch off at the end. To sense the record size, a second arm was fitted to the immediate left of the arm that carried the pickup. This arm was fitted with a lamp and a photocell and was able to distinguish between the smooth surface of a record and the black ribs of the platter. The two arms were styled to suggest a tuning fork, the reference of musical precision.

As all the controls were electronic, the Beogram 4000 did not need knobs or levers. This allowed the smooth, sleek design of the top surface to be realized, with all controls being flush whilst at rest. As well as controls for on, off and manual speed selection (for

Inside the Beogram 4000.

The control panel of the Beogram 4000.

non-standard records), the Beogram had a four-way control pad to cue the arm to any position. This was accompanied by an illuminated scale so that notes could be made on the record sleeves detailing where each track started. The control itself took the form of a square plate that rocked when pressed and was marked with arrows pointing north, east, south and west. North and south lifted and lowered the arm, while east and west traversed it across the record. The electronic logic ensured that the arm automatically lifted first before it was moved sideways. This form of control proved popular and was revived by Philips about ten years later for one of their very successful early compact disc players.

Given its technology and its content, it was not surprising that the Beogram 4000 was expensive. It cost

The Beogram 4002 combined simplified electronics and slightly revised styling to produce a new model.
Availability ●●●●
Complexity ●●●

1970 TO 1979: A DECADE OF ACHIEVEMENT 67

around a third more that the Linn Sondek LP12 in Britain, although the Linn did not include a pickup cartridge and needed careful setting-up before it could be used. It was also only a manual model that required considerable skill on the part of the operator if it was to give its best. In contrast, the fully automatic and easy to use Beogram 4000 was supplied complete and ready to run. It could be prepared in less than an hour and even came with a fitted mains plug.

The Beogram 4000 was later replaced by the similar looking, but greatly simplified, Beogram 4002. Somehow this managed to give similar facilities and performance to the 4000 but with about half the content – a great achievement. This was made possible partly by removing irrelevant or useless refinements (such as the bi-directional correction of the tracking angle included in the 4000) and partly by altering the arm-control system so that the digital electronic control unit was no longer needed. Later versions even dispensed with the electronic AC motor, using a simple DC servo unit sourced from Matsushita of Japan

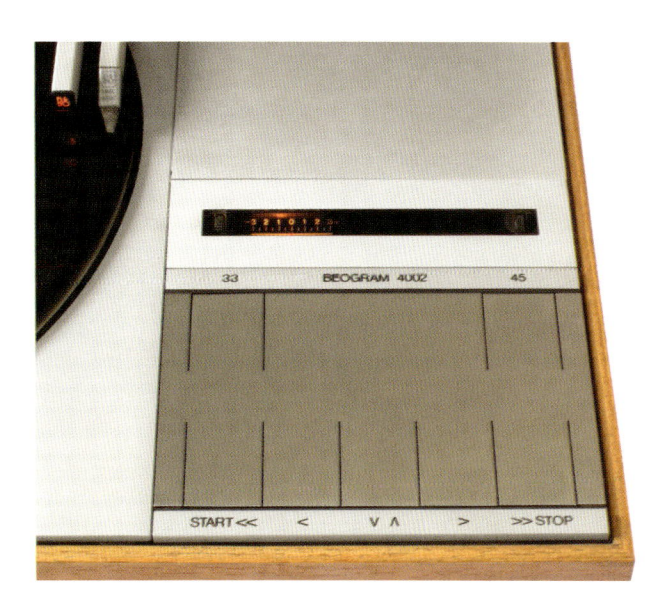

The new control panel matched those of other top models such as the Beocord 5000 and the Beomaster 6000 4channel.

A Beogram 6000, complete with a JVC CD4 quadraphonic record. This model was used as the source for the Beomaster 6000 4channel receiver.
Availability ●
Complexity ●●●●

instead. The Beogram 4002 also formed the basis for the quadraphonic Beogram 6000, the difference being the fitting of a CD4 decoder and the special MMC 6000 pickup. The previously mentioned Beogram 4004 was also based on the Beogram 4002.

The Beogram 4000 series is a collector's favourite. Good quality, fault-free examples are always in strong demand and are priced accordingly. View less than perfect models with suspicion, as the complex electronics and precision mechanical parts are very difficult to restore properly, and scratched, cracked lids, corroded aluminium and tatty woodwork spoil the sleek looks. Replacement pickup units are available from B&O dealers but, at the time of writing, their appearance does not match that of the originals and so, again, they can spoil the look of the machine as a whole. Damaged and faulty examples can be lovingly restored but do not underestimate the potential difficulty (or cost) of the task.

Before attempting to transport a Beogram 4000, even for a short distance, be sure to screw down the suspension, restrain the arm and remove the platter/centre spindle. Failure to do so can result in irreparable damage.

Other Turntables

Though the Beogram 4000 was a headline grabber, B&O carried on with the production of conventional radial turntables throughout the 1970s as well. The Beogram 1202 was the first important new radial model and although this looked similar to the previous 1200, it included a new (and vastly superior) suspension system and was a very strong performer. This model was also available as the Beogram 3000, which had a hinged dust cover and a better platter.

The basic manual range struggled on for a few years at the beginning of the decade, the last model being the Beogram 1001. In simple terms this was a mildly restyled Beogram 1000 with a new motor and only two speeds, as the 78 setting had been removed. Intended for use with the Beomaster 1001 in a budget entry-level system, it was not popular and is seldom encountered today.

By 1975 the Beogram 1202 and 3000 had spawned a bewildering array of variants, which tended to become simpler and more cheaply made as the years passed. The next important new model was the Beogram 1100, a slimline machine that was the first

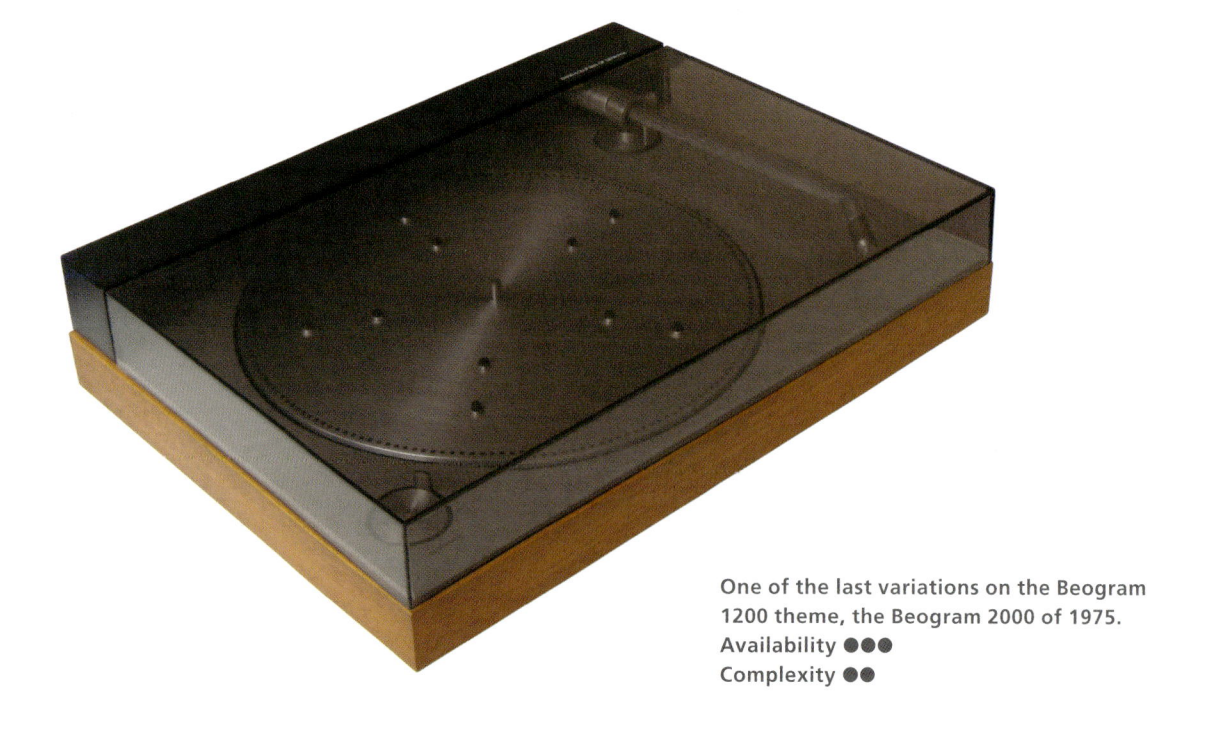

One of the last variations on the Beogram 1200 theme, the Beogram 2000 of 1975.
Availability ●●●
Complexity ●●

1970 TO 1979: A DECADE OF ACHIEVEMENT

radial Beogram to use the MMC range of pickups. The MMC 3000 version was made for these simpler decks and featured a spherical diamond stylus. The Beogram 1100 was also offered in a slightly improved specification labelled as Beogram 1900 to be sold with the Beomaster 1900. A quadraphonic version, known as the Beogram 3400, was also briefly produced for use with the Beomaster 3400. The CD4 decoder was not fitted in the factory and had to be specified separately. The MMC 5000 pickup was specially produced for this model.

The Beogram 1100 range would be the last new design to use an AC motor that was directly powered from the mains. An outwardly similar range soon

ABOVE: When the simpler models in the Beogram range were fitted with MMC pickups, they often used the MMC 20 S. The two most basic pickups in the MMC range were coloured black, the two more elaborate ones were finished in brushed metal.

ABOVE: The MMC 20 EN was fitted as standard only to the better classes of Beogram models.

BELOW: A good choice for both the collector and the keen listener alike, the Beogram 1700 is sturdy, reliable and capable of producing superb quality sound.
Availability ●●●●
Complexity ●●

ABOVE: The arm of the Beogram 1700 was so slender that it even managed to make the tiny MMC pickup appear large.

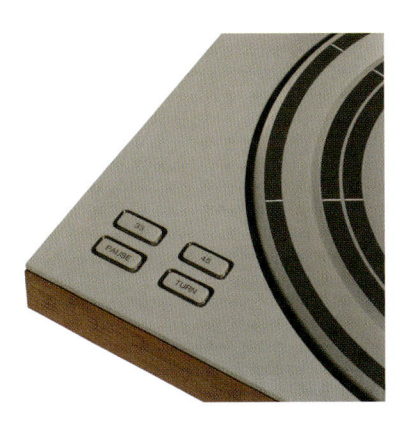

LEFT: These electronic control keys replaced the mechanical rocker plate of the earlier radial models.

replaced it that used the same DC servo motor that was fitted to the later Beogram 4002, the most common variant of which was the Beogram 1902. For the more financially constrained, the budget Beogram 1500 and 1102 models offered a more basic package, although the differences are difficult to spot at a glance. All these models can be easily recognized by the simple 'rocker plate' control surface in the front left-hand corner, which looks sleek but in practice has a rather heavy mechanical action.

Eventually electronic controls would appear as a feature of the radial models, initially on the Beogram 2200 and 2400. These models featured an angled lid and were clearly styled to match the Beomaster 1900 and 2400. Later, the very popular Beogram 1700 included them too; this is an excellent and very practical machine.

The First Cassette Recorders

The now familiar and universal compact cassette tape-recording format had first been shown by Philips in 1963. Originally intended as a format for simple, portable, speech-recording machines, the convenience and ease of use that the compact cassette system offered, transformed tape recording from a

RIGHT: The Philips pocket recorder of 1963 introduced the compact cassette tape format.

BELOW: The Beocord 1600 was the last open-reel tape recorder that B&O would market.
Availability ●●
Complexity ●●●

BELOW RIGHT: This switch bank looks impressive but in practice the Beocord 1200 and 1600 were nowhere near as versatile as the previous 2400 and 2000 De Luxe models.

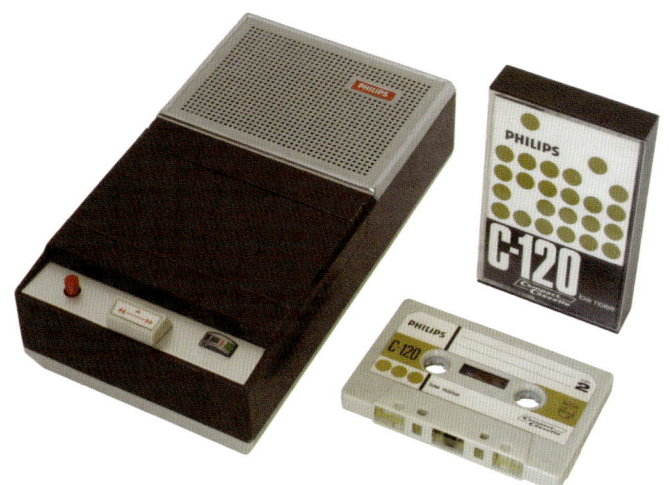

minority interest into a popular activity. Philips had broadened the cassette's appeal by releasing a wide choice of pre-recorded material along with ranges of more ambitious recorders and players, including some stereo models.

B&O entered the 1970s with a range of quality open-reel recorders but no cassette machines. The Beocord 1800 and 2400 were replaced by the Beocord 1200 and 1600, which were slim and attractive and followed the fashions of the day by being operable in the vertical position. They lacked the sophisticated facilities of the previous ranges, fitting more into the quality domestic bracket rather than the semi-professional class.

The first cassette model was the Beocord 900, which appeared in 1973. This stereo machine was made deliberately basic, as the emphasis was placed instead on ease of use. It did not offer the last word in quality but, if one ignored the rather poor 'chrome' setting and concentrated on ferric tapes, it could give quite acceptable results.

The next B&O cassette recorders were a more serious proposition. The Beocord 1700 and 2200, both of which appeared early in 1974, were serious high-end models with performance potential to match. They were fitted with wear-proof ferrite heads and an electronically regulated motor, both of which, along with the cassette mechanism, were sourced from Japan. The electronics were all B&O's work, however, and included every possible adjustment, so that the machine could be matched to whichever chrome or ferric tapes the owner chose.

The Beocord 1700 was sized and styled to match the Beolab 1700 audio system. This could be arranged in a 'stack' – a first for B&O. The Beocord 2200 looked identical but was fitted with an important addition: Dolby B noise reduction. It is the Dolby process that is credited with transforming the compact cassette into a proper high-fidelity medium. By removing much of the 'hiss' associated with tape recording, the music could come through uncorrupted, and carefully made recordings could match the sound of the best LPs.

The Dolby circuit of the Beocord 2200 was very much a first-generation affair and so was constructed from many discrete components. Later generations would benefit from complete Dolby circuits fabricated on a single silicon chip. Beocord 1700 owners could buy the Dolby module separately and have it fitted to their machines by a dealer.

As Dolby noise reduction became a necessary part of any stereo cassette deck, the Beocord 900 was redesigned to include this feature. The new model was known as the Beocord 1100 and could be recognized by its black fascia (the Beocord 900's was natural aluminium). A special model was also produced to match Beomasters and Beocenters with natural aluminium finishes. This was known as the Beocord 1101.

The Beocord 2200 was B&O's first high-fidelity cassette deck. It was amongst the first on the market to include the famous Dolby B noise reduction system.
Availability ●●●●
Complexity ●●

Impressive though the Beocord 2200 had been, B&O's engineers clearly thought they could do better. The next model represented a massive investment in time and resources, and aimed to be the best cassette recorder the world had ever seen. Known as the Beocord 5000 when it was released in early 1976, the new machine was styled along the lines of the Beomaster 6000 4channel, launched a few years previously.

Although not capable of producing four-channel recordings, the new recorder matched the receiver well and continued the theme of electronic operation and illuminated display panels. Internally it was massively complex. For the first and only time, B&O produced their own cassette mechanism, which was unlike anything that had been seen before. It used two servo-controlled motors and two capstans in an attempt to ensure absolute speed stability. The capstan servo monitored the actual tape speed instead of the motor speed, so inconsistencies in the belt drive were automatically cancelled. The reel servo employed an unusual electronic technique known as 'negative resistance' to give precisely controlled

ABOVE: **The Beocord 1100 was easy to operate and matched a wide range of Beomaster and Beocenter models.**
Availability ●●●
Complexity ●●●

ABOVE: **The tape transport keys were made of anodized aluminium, the same material as the rest of the front of the machine.**

LEFT: **Open and ready to receive a cassette. Convenience and ease of use were major advantages of cassette recorders over open-reel machines.**

1970 TO 1979: A DECADE OF ACHIEVEMENT **73**

RIGHT: The electronic controls of the Beocord 5000 gave it a sleek appearance. No other cassette recorder of the period looked so elegant.
Availability ●●●●○
Complexity ●●●●●

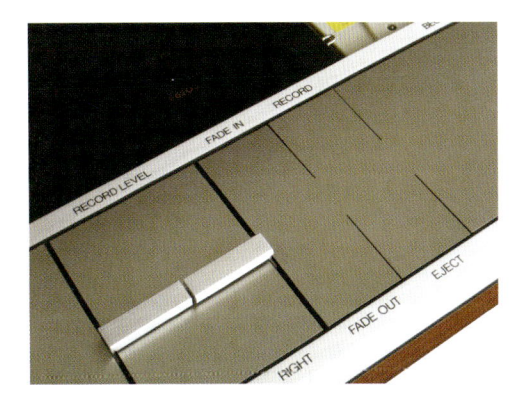

The controls matched those of the Beomaster 6000 4channel. Note that there are two stop buttons to keep the styling neat.

These sliders controlled the recording level.

tension on the tape. All the deck functions were controlled by solenoids and the operation of the various functions was sequenced automatically. Electronically, the Beocord 5000 boasted Dolby B noise reduction, a microphone amplifier, automatic fade-out and fade-in functions, and separate recording and replay heads, both made from wear-proof ferrite.

The separate heads could have been used to allow 'off the tape' monitoring (as was possible with the Beocord 2000 De Luxe) but, unfortunately, they were positioned with the recording head following the playback one, making this feature impossible to implement. B&O's cassette mechanism required a large number of precise mechanical adjustments to be maintained if it was to function correctly. This proved difficult to do in practice and the Beocord 5000 soon gained a reputation for being temperamental. This original model was known as type 4705; it was soon replaced with type 4715. This second version omitted the separate recording and playback heads, and instead a combined record/replay head made from Sendust was used. It was claimed that this new layout could give more consistent results and was easier to match to tapes. This was true to an extent but the Beocord 5000 was still a challenge to keep on top form. When working it was capable of outstanding results, however, and it remained in the catalogue until 1979.

Due to their complexity, few working examples of either version still exist today, with type 4705 being particularly rare in operable condition. They are best viewed as an interesting addition to a collection, rather than a useful everyday recorder; although the simpler models are still capable of good service. Cassette recorders improved greatly in the 1980s, so for making a large archive of recordings, a later model is recommended, although of course the tapes can still be played on older machines.

Music Centres

The music centre is the style of audio equipment most commonly associated with the 1970s. Compact transistorized circuitry meant that it was no longer necessary to have a large radiogram cabinet to house a complete integrated audio system, especially as the public had at last recognized that better sound reproduction could be obtained from two well-positioned separate loudspeaker enclosures.

Unfortunately, manufacturers tended to place music centre models at the bottom of their ranges and equipped them accordingly; this gave the music centre the reputation of being the 'poor relation', although this was often richly deserved.

B&O's first music centre was different. The Beocenter 3500 of 1972 was made by combining two top-line components: the Beomaster 3000-2 receiver and its matching Beogram 3000 turntable into one neatly styled and beautifully finished cabinet. All of the Beocenter 3500's front panel and top surface gleamed with finely machined aluminium and the advanced circuitry inside gave it a performance advantage no competitor could match. The Beocenter 3500 was very expensive but it was a strong seller; it continued in the range until 1977. Later examples are easily identified as the front panel was made black instead of silver and the hairlines in the slide cursors were green instead of orange.

1973 saw a new type of music centre from B&O. The Beocenter 1400 combined the receiver chassis of the Beomaster 901 and a simplified cassette deck based on the Beocord 900. A record player, such as the Beogram 1202, could be connected as well, although the Beocenter 1400 was nevertheless complete as it stood and represented a thoroughly up-to-date concept in the early 1970s. In 1976 the Beocenter 1500 replaced the 1400 model, the main change being that the tape transport mechanism was now the improved type as used in the Beocord 1100, although Dolby NR was still not fitted. An FM-only version based on the Beomaster 1100 was also introduced; this was known as the Beocenter 1600 and came with an elegant black fascia. This series of models are amongst Jacob Jensen's most visually

All the exposed parts of the Beocenter 3500 were made from solid metal. The build quality is of a very high standard.
Availability ●●●●
Complexity ●●

1970 TO 1979: A DECADE OF ACHIEVEMENT   75

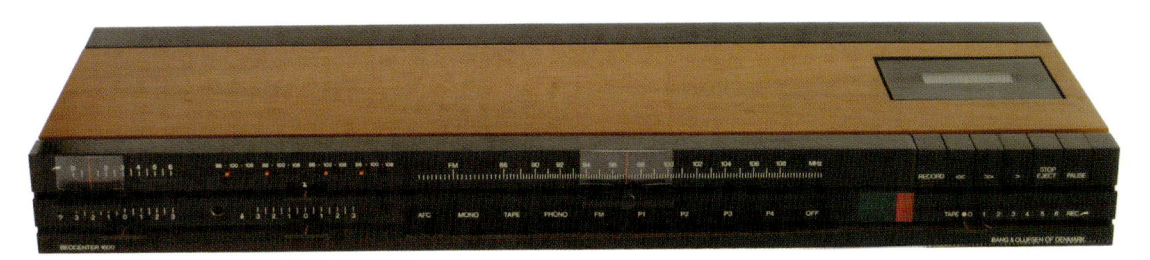

Based on the Beomaster 1100, this Beocenter 1600 is a classic example of the B&O cassette music centre.
Availability ●●
Complexity ●●

ABOVE: The comprehensive amplifier controls of the Beocenter 1600 were identical to those of the Beomaster 1100 receiver…

…whereas the cassette deck used the same mechanism as the Beocord 1100.

The Beocenter 1800 used the latest pickup technology in a compact, high-quality unit that was easy to understand and operate.
Availability ●●
Complexity ●●

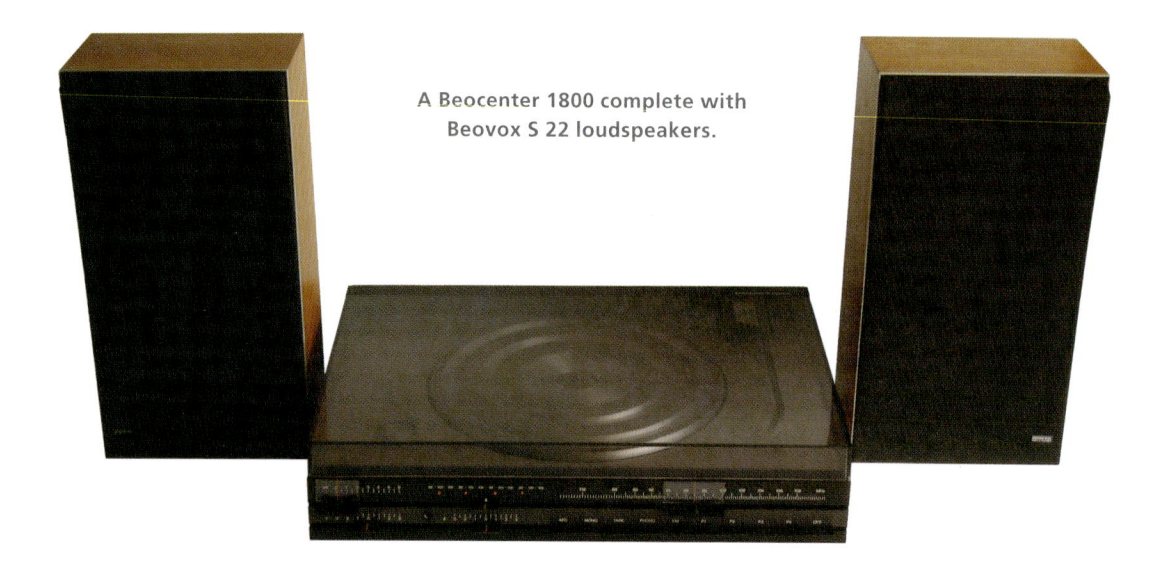

A Beocenter 1800 complete with Beovox S 22 loudspeakers.

The turntable of the Beocenter 1800 was completely automatic in operation. The smaller dial was used to fine-tune the speed.

successful, as the extra length added by the inclusion of the cassette section gives a stunning impression of sleek elegance. They are thoroughly recommended as part of any collection.

It was still expected by many that a music centre should include a turntable, so to cater for those with this requirement, who found the Beocenter 3500 either too big or too expensive, the Beocenter 1800 was introduced. This combined the Beomaster 1100 and the Beogram 1100, and formed a very neat and attractive unit. This was the first music centre to come fitted with an integrated MMC pickup cartridge.

By the mid-1970s some manufacturers had started producing music centres that combined a radio, a turntable and a cassette recorder. B&O's range up to this point had included two of these sources with the option of adding the third externally, although it was recognized that this was not always convenient. To fill this gap, the Beocenter 3600 appeared in late 1976. This model combined the Beomaster 1100, the Beocord 1100 and the Beogram 1100 into one unit. Unfortunately the styling was not well-executed and this, along with certain technical problems, resulted in this model not being very popular. The follow-up Beocenter 4600 of 1978 was altogether a better piece of work and this sold very well.

The Beocenter 4600 was part of a new generation of music centres that were based on the Beomaster 1500 and 2200 receivers. These new high-performance models kept the Beocenter range at the forefront of technology and performance. The Beocenter 4600 was available without the cassette deck as the Beocenter 2800 and without the turntable as the Beocenter 2600.

The Beomaster 2200-based range comprised two models. The first of these was the Beocenter 3300, which combined a slightly shortened Beomaster 2200 with a Beogram 1902-type record deck. Styling needs

1970 TO 1979: A DECADE OF ACHIEVEMENT 77

The Beocenter 3600 was a very up to date concept but did not sell in large numbers.
Availability ●
Complexity ●●●

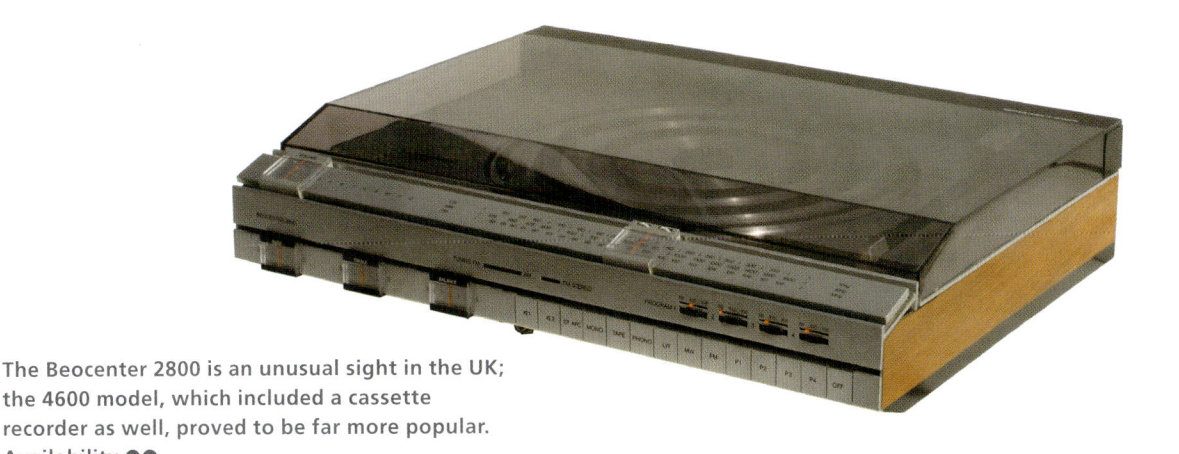

The Beocenter 2800 is an unusual sight in the UK; the 4600 model, which included a cassette recorder as well, proved to be far more popular.
Availability ●●
Complexity ●●

dictated that the turntable was placed next to the receiver, rather than above it, resulting in one of the widest audio units B&O had ever produced.

The second model was known as the Beocenter 4000 and this was rather special. The concept of a music centre based around cassettes, rather than records, was repeated but this time the cassette mechanism chosen was a simplified form of that used in the high-performance Beocord 5000. The process of simplification included the removal of one of the capstan and pinch roller pairs and this transformed the machine from a temperamental device that occasionally ruined tapes into a really super machine that finally lived up to expectations. This alone recommends the Beocenter 4000 for any serious B&O collection, as it is the most successful implementation of the only cassette mechanism that can truly be called a B&O original.

The electronic tape transport controls of the Beocenter 4000 were styled to match the source and station selectors of the Beomaster and the cassette compartment was concealed under a hinged flap that complemented the one that covered the minor amplifier controls. This model was by far the best of all the

1970 TO 1979: A DECADE OF ACHIEVEMENT

A music centre for the serious cassette user. Fitting a high-performance cassette deck into the Beocenter 4000 created a unique product whose performance was outstanding.
Availability ●●
Complexity ●●●●

The minor controls and the cassette mechanism of the Beocenter 4000 were hidden under two aluminium covers.

A close-up view of the amplifier controls under the left-hand lid. Also visible is the pop-up bank of pre-set tuning controls.

cassette receivers (although inside it was very complicated) and certain elements of the styling were later revived in the Beocenter 9000.

Music centres offer an ideal way to start or to build up a B&O collection. The Beocenter 3500 is an excellent choice and needs much less space than the equivalent Beomaster and Beogram. It is also an outstanding performer and easily expanded into a full system. Of all the various models, however, it is the cassette receivers that represent the best value. Compact and easy to use, they can form a complete audio system without the problems of record-playing mechanisms and the expense of pickup cartridges, along with the real advantage that you can make your own recordings and so listen to whatever you like, without having to resort to additional or non-period equipment. Although not the best equipped, the Beocenter 1500 and 1600 are perhaps the models of choice. They are simple enough not to suffer from serious maintenance worries, remarkably compact and show Jacob Jensen's styling at perhaps its most elegant. At the time of writing, they are also inexpensive to buy and not too difficult to find. At the opposite end of the spectrum, the Beocenter 4000, due to its complexity, is best approached with care by the inexperienced, if not in perfect working order. It is not uncommon to encounter Beocenter 4000s with the covers over the amplifier controls and cassette compartment broken.

Repairing these is difficult but not impossible, as replacement parts can be fabricated, if necessary.

Loudspeakers – the New Phase

B&O had always offered a range of quality loudspeakers to match their equipment. Although suitable types were suggested to go with each model, the final choice was usually left up to the customer, allowing the constraints of room layout and décor to be fully accommodated. From the point where the company became a major exporter in the 1960s, up to the middle of the 1970s, the Beovox range of loudspeakers had been well-finished and technically competent, although in simple terms most of them were simply square wooden boxes with black cloth fronts containing conventional drive units operated in the usual manner.

Having introduced new and world-leading technologies in most areas of home entertainment, it was only a matter of time before a more interesting range of loudspeakers emerged.

1976 saw the introduction of a new loudspeaker line known as Uniphase. Still fairly conventional in appearance, these new models used new techniques to improve the phase linearity of the loudspeaker system, to ensure that the conversion process from electricity to sound took the same time, regardless of frequency. This helped to make the illusion of a solid sound image seem all the more convincing. Uniphase loudspeakers could all be recognized by their angled front panels, leaning out towards the top of the baffle. It was claimed that this design moved the tweeter marginally closer to the listener, helping to make up for the 'lag' introduced in the crossover network.

Although the angled baffle was outwardly the most obvious part of the Uniphase concept, it was only a small part of a larger picture. New, more efficient drive units and more powerful amplifiers meant that the designers could be more creative with the design of the crossover circuits, freed from the necessity of the pursuit of efficiency at all costs. The smallest amplifier in the range in 1976 was the 20W design fitted to the

ABOVE: The Beovox 5700 was the top of the range Beovox loudspeaker in the early 1970s.
Availability ●
Complexity ●●

ABOVE: The conservative styling hid the most up to date drive unit technology, such as the two dome units visible in the centre of the baffle.

BELOW: The angled baffle of the Beovox Uniphase loudspeaker range is most clearly seen in this profile view of a later S 80.2 model.

Beocenter 1500, which was still easily powerful enough to drive quite sophisticated loudspeakers.

The original Uniphase range comprised, in order of size, the S 30, the S 45, the S 60 and the M 70, the latter being a large four-way monitor-grade design sold complete with a steel floor stand. 1977 saw the miniature S 22 model added to the range, unusual at the time because it used drive units from Japan, which have proved to be very durable.

The whole range was revised in 1978 – the new models being denoted either with a larger number (e.g. the S 60 was replaced by the S 75) or by adding a '.2' to the end, as in the S 45.2. The S 45.2 was a particularly fine loudspeaker and is a worthy addition to any mid-range 1970s system. At the same time, the range was expanded upwards with the colossal five-way M 100 model. The M 100 was very complicated as some of the drive units were used as 'phase links' to smooth the response of the system overall. In the following years, the approach of using so many different drive units would be abandoned, as it was very difficult to do properly and was made unnecessary by advances in loudspeaker design. At the other end of the range, the S 25 (which replaced the S 22) took a move downmarket, as the dome tweeter was dropped in favour of a simpler, cheaper, cone type.

The Uniphase principle was also applied to the range of wall-mounted panel loudspeakers, the best known of which were the P 30 and the P 45. The P 45 was roughly the equal of the S 45 and again cannot be recommended highly enough.

The Uniphase concept was soon applied to a range of miniature loudspeakers housed in die-cast aluminium cabinets. The original range comprised of the Beovox C 30, C 40 and C 75, the former two models having two drive units and the latter having three.

The original models were complex and included internal porting through a grille at the rear to improve the bass performance. The original 'C' series models were later replaced by the hugely successful CX 50 and CX 100. These would remain available for the next twenty years and were ideally suited to multi-

ABOVE: **Uniphase in miniature: the Beovox Uniphase S 22. The curved baffle and shaped moulded rear panel can be clearly seen in this picture.**
Availability ●●
Complexity ●

RIGHT: **The slim profile of the Beovox Uniphase P 30 cabinet can be seen here.**

ABOVE: **Beovox Uniphase P 30 wall-mounted loudspeakers.**
Availability ●●●
Complexity ●

The amazingly compact Beovox C 30 is the smallest complete passive loudspeaker that B&O have produced.
Availability ●●
Complexity ●●

When viewed from the rear, it can be seen that the Beovox C 30 (on the left) has a port at the rear and the Beovox CX 50 (on the right) does not. The Beovox CX 50 has the same size cabinet as the original C 40 model.

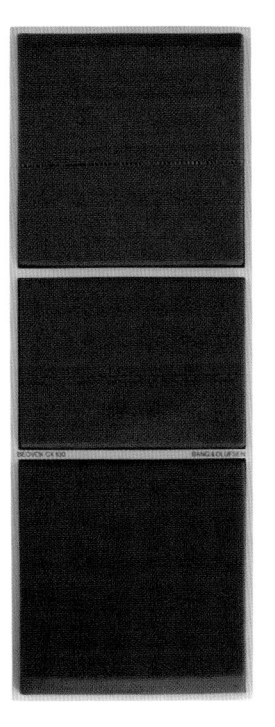

LEFT: A Beovox CX 100, viewed directly from the front. The whole cabinet is no taller than an LP record in its sleeve.
Availability ●●●●●
Complexity ●●

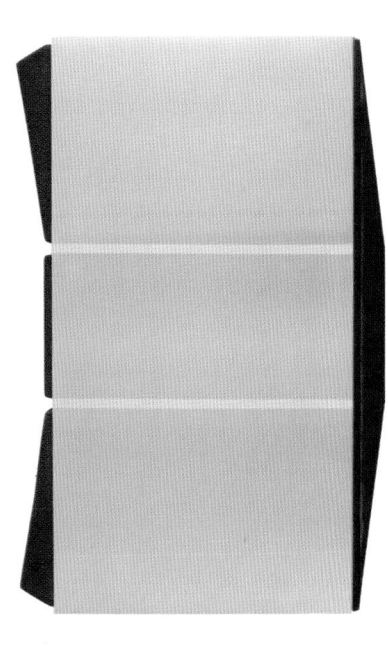

ABOVE: The Beovox CX 50, which replaced the original C 40 model
Availability ●●●●●
Complexity ●●

LEFT: From the side, the Beovox CX 100 bears all the hallmarks of a Uniphase design.

room setups. The construction of the CX models was slightly simpler than their previous C series equivalents, with slightly smaller drive units and fully sealed cabinets.

The original range of Uniphase loudspeakers were made before foam rubber roll edges were the norm and so buying them does not, in all cases, leave one worrying about expensive or difficult repairs. However, the serious listener should pay close attention to the capacitors in the crossover networks. These were originally electrolytic types, not of the best quality, and the passing of thirty or more years has not improved them. Modern high-quality alternatives can work wonders with a tired sounding pair of loudspeakers.

All Beovox C and CX loudspeakers have foam roll edges and all will rot away given time. Inspect any potential purchase very carefully, even if the seller claims that the particular pair is not very old, as the first sets produced look very much like the last.

Progress in Television

By 1970, colour television had become firmly established across Europe. The Beovision 3000 was updated by the fitting of a slightly larger, flatter and squarer picture tube, of a type that could be 'pushed through' the front of the set, rather than framed in a mask. This model was known as the Beovision 3200 and in the following year, a smaller set using the same

ABOVE RIGHT: Despite the smaller screen, the Beovision 2600 was still a bulky set.

ABOVE LEFT: The Beovision 2600 was one of the last models to use B&O's first-generation colour chassis, which had first been seen in the Beovision 3000.
Availability ●
Complexity ●●●●●

BELOW LEFT: The minor controls were concealed in three pop-out drawers. The control to the bottom-right is a 'tint' control, which alters the balance of the colours.

ABOVE: **The bulk of the Beovision 3400 had been reduced so much that the screen filled nearly all the front of the cabinet. This was a great advance over the previous generation of colour receivers.**
Availability ●
Complexity ●●●●●

ABOVE RIGHT: **This view shows how slim the Beovision 3400 was. Compare this with the Beovision 3000 shown in the previous chapter.**

BELOW RIGHT: **This hidden drawer contains some of the complicated adjustments that were needed to set up the wide-angle tube of the Beovision 3400.**

chassis but fitted with a smaller 22in tube was also introduced, this was known as the Beovision 2600.

B&O had always worked hard to make their equipment blend well in domestic surroundings and one thing that made their early colour TV sets difficult to house was the depth of their cabinets. The Beovision 2600 was large and the 3200 truly vast, mostly due to the depth of the picture tubes. Some relief came in the early 1970s when Philips announced the Phase 2 Delta wide-angle colour tube that had the same screen size as the existing types but was appreciably shallower. The trade-off was that this tube was more difficult to operate and required far more scanning power. B&O were amongst the first to adopt the Phase 2 Delta tube and by the middle of 1972 had released the Beovision 3400, a slim colour set with a 26in screen.

The Beovision 3400 was available in two cabinet styles: one a traditional set, with legs and a roller door over the screen; and the other, a remarkably compact table model. A 22in-version of the set, the 3100, was later released but this did not prove very popular as the size of the chassis meant that the cabinet was no smaller, giving the set a rather awkward appearance.

Although visually similar to the Beovision 3000 models internally, the Beovision 3400 was really a completely new set. Advanced techniques were required to produce enough scanning power for the wide-angle tube and elaborate circuits had to be designed to perform the multitude of geometric corrections that were also needed. In some cases, transistors replaced valves, although ten of the previous eighteen valves remained. One particular problem was that of power dissipation. The Beovision 3000 and its derivatives had a reputation as being hot running sets and the 3400 would need even more power to drive the tube. The cabinets were also smaller, creating a real problem in terms of internal temperature.

One of the most power-hungry parts of the 3000 design had been the high-voltage generator. This was made highly stable at the price of efficiency to ensure that the picture remained exactly the same size under all possible conditions. Burning up this much power in the 3400 was not acceptable, so a new and very clever solution was employed. The high voltage was only to be very loosely regulated to keep it within the bounds that, at the low end, yielded adequate focusing and, at maximum voltage, minimal forward radiation of X-rays from the screen. The resulting voltage would then be measured and the picture instantaneously resized, refocused and corrected in brightness to suit. As unlikely as it seemed, this technique worked perfectly and the Beovision 3400 was capable of performance every bit as good as that offered by the previous models.

ABOVE: The Beovision 601 was the last monochrome Beovision. Special circuits meant that a perfect picture appeared the moment the set was turned on.
Availability ●●
Complexity ●●●●

ABOVE RIGHT: The controls of the Beovision 601 were concealed in drawers, just like those of the larger colour sets.

BELOW RIGHT: The metal strip around the cabinets of the Beovision 600 and 601 slides up to become a carrying handle.

1970 TO 1979: A DECADE OF ACHIEVEMENT

At launch, the Beovision 3400 was undoubtedly the most complex consumer product ever offered to the public. Initially there were problems with reliability that got so bad that B&O issued their dealers with detailed inspection and overhaul instructions that were to be followed whenever a set was brought in for service. Some of the components that were originally fitted also proved inadequate and so B&O offered a pack of replacement items free of charge to be fitted at the same time. Despite its advanced technology, the Beovision 3400's reliance on valves and its (admittedly excellent) mechanical tuner made it, in some ways, out of date even when launched. This would be rectified in the next colour chassis design.

Meanwhile in the monochrome line, a fully transistorized model with an electronic tuner had been introduced. Known as the Beovision 600, this 15in portable set was of a very modern design. The transistorized electronics came to life the instant the switch was pressed, so the designers arranged the circuit to make the picture appear instantly. A picture tube, like a valve, has parts of its internal structure that must be heated to a very high temperature before it will work. This normally takes a few seconds, but to avoid even this delay the Beovision 600 had its tube heater running at a reduced level of power, even when the set was switched off. This produced a perfect picture the instant the switch was pressed, a feat that has not been repeated since. Instant picture aside, the Beovision 600 was a neat and attractive set. The mains cable recoiled inside when not in use and the sound, picture and tuning functions were all controlled by little wheels hidden behind pop-out drawers.

The same chassis was fitted to B&O's last large-screen monochrome set, the Beovision 1600. As well as a larger screen, this model also included separate treble and bass controls for the sound. In 1973, the chassis of the Beovision 600 was updated to use an integrated circuit in place of a large number of separate components in the signal processing stages and was renamed the Beovision 601. This model would be the last monochrome set that B&O produced. It was also the last portable television for which B&O designed both the chassis and the cabinet.

Transistorized Colour TV

With 1974 at last came the first fully transistorized colour sets. Produced in a wide range of cabinets and screen sizes, this series was epitomized by the Beovision 3500 and Beovision 6000 models. The Beovision 3500 was a remarkably compact and beautifully styled 22in screen model finished in gleaming white. It made full use of the possibilities offered by the new chassis and the Philips wide-angle tube. It was observed that the set was no larger and no heavier than a similarly sized monochrome model of a few years previously and that it could be carried easily by one man.

In a few short years the colour television had evolved from a temperamental beast housed in a bulky piece of furniture into a compact and reliable product like this Beovision 3500. The pre-eminence of European design in this period is clearly visible in its beautiful lines and fine detailing.
Availability ●●
Complexity ●●●●

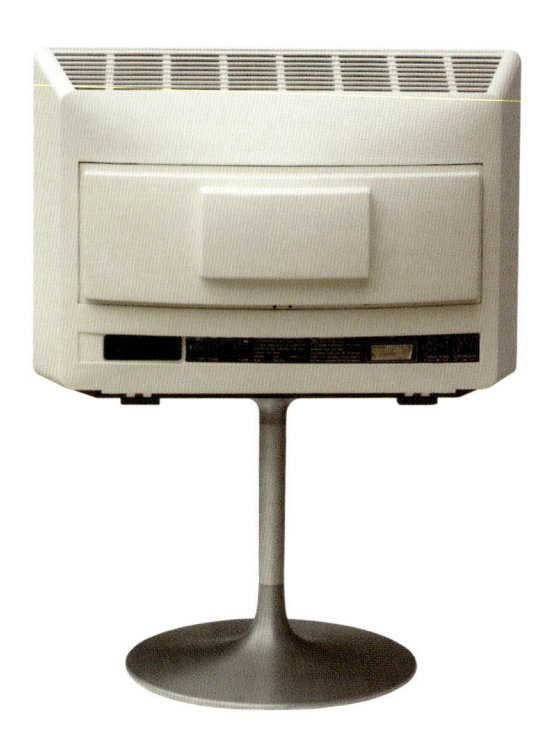

ABOVE: Concealing the minor controls in pop-out doors keeps the control panel sleek and easy to use. There is also no 'on' button – selecting a channel to watch turns the set on automatically.

RIGHT: A rear view of the Beovision 3500 shows how advancing technology had eliminated the need for pre-set control spindles and copious ventilation grilles.

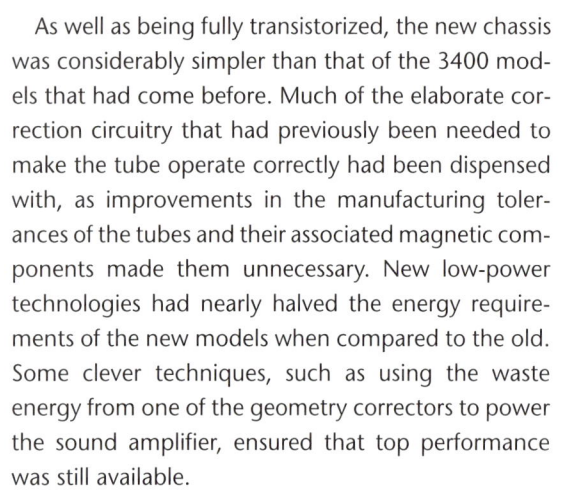

As well as being fully transistorized, the new chassis was considerably simpler than that of the 3400 models that had come before. Much of the elaborate correction circuitry that had previously been needed to make the tube operate correctly had been dispensed with, as improvements in the manufacturing tolerances of the tubes and their associated magnetic components made them unnecessary. New low-power technologies had nearly halved the energy requirements of the new models when compared to the old. Some clever techniques, such as using the waste energy from one of the geometry correctors to power the sound amplifier, ensured that top performance was still available.

During this period, television design was advancing at possibly the fastest rate ever. The new Beovisions included an important innovation known as Permanent Colour Truth. In previous models, the light output of each of the three primary colours that make up the colour picture had to be carefully adjusted to give correct results. As the tube and its driving components aged, these adjustments would slowly drift off giving incorrect colours on the screen. Only the most quality-conscious owners had their sets adjusted regularly, so the majority of colour sets were normally not showing the correct hues. The new Beovisions re-adjusted themselves fifty times a second and so never drifted off or showed the wrong colours, at least not until the picture tube was completely exhausted. This system became a key part of every cathode ray tube Beovision set that followed and was also adopted by other manufacturers.

The Beovision 6000 was placed at the very top of the range. It was a 26in set in a large cabinet, similar in size and style to that of the previous 3400. The new feature of the 6000 was remote control. This was a first for B&O but it had been available on other makers' sets for some time. Ultrasonic sound was used to send the commands from the remote control unit to the set. This was common practice at the time but in its normal implementation the system suffered certain drawbacks, notably that jangling keys, ringing doorbells or telephones and even certain voices would trigger the system and make the television behave erratically. B&O solved this problem by making the remote control unit transmit two sound frequencies simultaneously for each function. It was highly unlikely that these two exact frequencies would occur together naturally and so the system gave perfect results under all conditions. The remote control unit, called the Beovision 6000 Commander, was similar to the one supplied for use with the Beomaster 6000 4channel quadraphonic audio system.

Advances in tube technology brought the next generation of models in 1977. The new Philips 20AX tube system had its three electron guns arranged in a line rather than in a triangle. This principle had been demonstrated by Sony with their Trinitron some years previously and since then various manufacturers, such as RCA and Toshiba, had produced in-line tubes of their own designs. A feature of the Philips tube was again its very modest depth and this enabled B&O to continue producing slimmer and slimmer sets. The 20AX tube needed so few correction adjustments that it was no longer necessary to have the set installed and adjusted by a dealer; the owner could take it home, tune it in and use it with no difficulty at all. This, combined with the now excellent reliability record of the sets of this period showed that, in ten short years, the colour television had evolved from a finicky and troublesome beast into the perfect consumer product. It had also become far more affordable and sales boomed. High-points in the 20AX-based range were the Beovision 3502 and 3802, which were similar in appearance to the previous 3500, the Beovision 4402, a top-selling elegant and refined 26in table model

The Beovision 3802 looked very similar to the 3500, but inside everything was new. Note the ultrasonic sensor at the bottom of the loudspeaker grille and the standby indicator set into the 'off' button.
Availability ●●●
Complexity ●●●

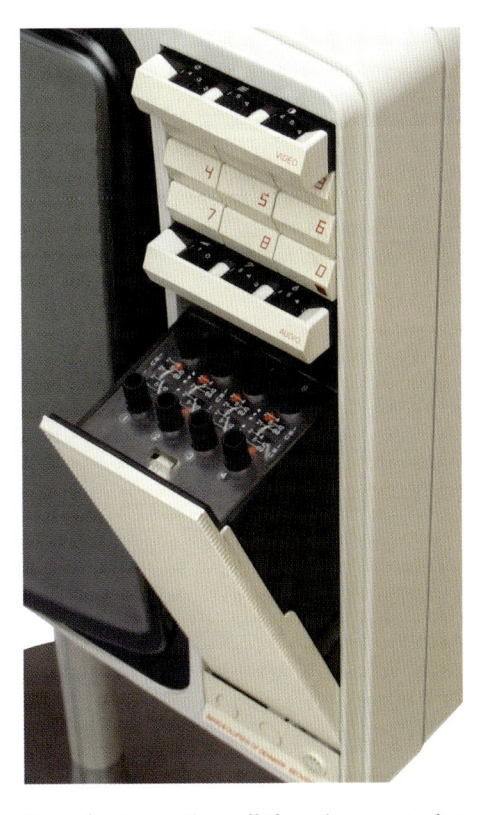

Open for inspection, all the minor controls can be seen. The faceted control keys indicate that this is one of the new '02' models fitted with the Philips 20AX tube.

1970 TO 1979: A DECADE OF ACHIEVEMENT

LEFT: The 26in Beovision 4402 was one of the most popular models in the series.
Availability ●●●
Complexity ●●●

RIGHT: The right-most of the three circles is a special microphone, which picks up the signals from the ultrasonic remote control unit.

with remote control, and the Beovision 6002, which replaced the previous 6000 and remained as a top-of-the-range cabinet set with every feature available, including B&O's first on-screen digital programme display. Rarest of this series is, undoubtedly, the 3502 in an optional aubergine finish.

The ultrasonic remote control handset supplied with these new sets had been made more compact and more user-friendly. The sharp edges of the previous model had been replaced with gently radiused curves and the buttons were made concave, convex or flat, depending on their function. Designed by Henning Moldenhawer and dubbed the 'Feel Commander', this new remote control could be accurately operated using only one hand without looking at it, truly a design masterpiece.

The 20AX-based Beovision models are the oldest sets that the collector, who is not also a television specialist, should consider. While they are still not guaranteed to be fault free, they have proved to be generally reliable in service and not too difficult to repair. Beovision 4402 or 3802 both make excellent sets that can easily be maintained in working order. Examples with tidy cabinets and strong tubes can still be found and, at the time of writing, can be bought inexpensively. The special battery that the remote control uses is no longer easily available but a 12V miniature type intended for car alarms can be adapted to fit. Remember that the remote control units use ultrasonic sound, rather than infra-red light, so a modern 'universal' remote control unit will not work. Ensure that a usable original is included with the set.

The new Feel Commander remote control had a gently radiused moulded plastic case that made it comfortable to hold and operate.

This view shows the concave and convex keys that made the Feel Commander easy to operate by touch alone.

CHAPTER FOUR

1980 TO 1989: THE MICROPROCESSOR AGE

The 1970s had treated B&O kindly and as the decade drew to a close they could offer a range of competitive, attractive models that would suit most tastes. Hi-fi and colour TV were enjoying what would later be recognized as a golden age of great public interest, a stability of formats and technological advancement.

As the 1980s progressed, more factors would play into B&O's hands. The Japanese Yen would strengthen, no longer making it easy for companies such as Denon, Technics and Sony to flood the European market with professional-quality equipment at high street prices. Video and home computers expanded the market for TV sets, as second and third sets were keenly acquired, and the compact disc would once again put hi-fi firmly in the spotlight.

Technologically, the big news of the early 1980s was the microprocessor. This was a device that had the computing power of a whole rack of equipment from only ten years previously, compressed onto one tiny silicon chip. It brought with it new possibilities of convenience and automation to all types of products, and mass production meant that it was cheap enough to be included in the better classes of domestic equipment. B&O were amongst the first to offer a full range of microprocessor-controlled models with new designs in all their product ranges.

Beolab 8000

It was the Beolab 8000 system that represented B&O's most distinctive and thorough entry into the microprocessor age. This three-part system (receiver, tape recorder and turntable) included sophisticated electronics in each unit and offered many new features and facilities.

The Beomaster 8000 receiver was the largest and most powerful receiver that B&O would produce. Its power amplifier was rated at 150W, a staggering figure that made it at least twice as powerful as any other Beomaster. As well as this, the Beomaster 8000 also

The complete Beolab 8000 system was very large but was capable of a level of performance that had not been seen before in such an easy to use audio system.
Availability ●●●
Complexity ●●●●

1980 TO 1989: THE MICROPROCESSOR AGE   91

RIGHT: The mighty Beomaster 8000 was the largest and most powerful stereo receiver that B&O ever produced.
Availability ●●●
Complexity ●●●●

BELOW: A beautiful juxtaposition of fine materials, quality finishes and high technology.

ABOVE: The minor controls for the tuner and amplifier were hidden under this lift-up panel.

ABOVE: Making connections to the Beomaster 8000 was easy as the sockets were housed under this special flap.

RIGHT: The Beolab Terminal was the first B&O remote control that could operate a whole system of separate components.

featured a microprocessor-controlled pre-amplifier and FM tuner. Settings for volume and balance could be set and stored electronically, along with up to nine FM radio stations. The main controls took the form of two large dials: one for volume and one for tuning. Both of these controls worked electronically and so had no mechanical end stops; the tuning control spun easily so that the FM band could be scanned quickly but the volume control was damped so that the massive power of the amplifier could be kept in check. The Beomaster 8000 came with infra-red remote control as a part of the standard package. The remote control unit worked not only the basic functions of the receiver but also those of the tape recorder and turntable. This was done, not by fitting infra-red detectors to all the units but by allowing them to communicate via a system that would become known as Datalink. Datalink was a standard for sharing control instructions between B&O audio units. Extra pins were fitted to the DIN plugs and sockets that interconnected each piece of the system, so that the instructions could be transferred without the need for extra wiring.

Datalink was a two-way system: this meant that the Beomaster receiver could control the sources and the sources could control the Beomaster. This made the operation of a complex system very straightforward. For example, to hear a record play, all one had to do was to press 'Play' on the turntable or 'PH' on the receiver; nothing else was necessary in order to hear the music, not even switching either unit on.

The Beocord 8000 tape deck was designed to match the lines of the Beomaster 8000. It too used a microprocessor to control its tape transport functions, which could of course be operated by remote control. Clever as this was, the great advance in the Beocord 8000 was the real-time tape counter. Previous tape decks had used mechanical counters linked to the

ABOVE: The cassette decks for the Beolab 8000 system were stylish and elegant machines. This is the later Beocord 8004 model, but the original Beocord 8000 looked practically identical.
Availability ●●●
Complexity ●●●

RIGHT: Microprocessor control meant that the main functions of the machine were controlled using a computer keypad.

1980 TO 1989: THE MICROPROCESSOR AGE   93

RIGHT: **Advanced technology made the Beogram 8000 as easy to operate as any record player, yet its sound quality was rivalled only by the very best.**
Availability ●●●
Complexity ●●●●

LEFT: **A close-up view of the arm and platter shows that they are similar in design to those used in the Beogram 4000. When a record is not playing, the arm retracts completely into the space on the right.**

ABOVE: **The control keys and the digital speed readout of the Beogram 8000.**

rotation of one of the spools to give a numerical index of tape position. This worked well enough but as the amount of tape changed on the spool that was being monitored, the rate that the counter numbers incremented changed too. There was no agreed standard for the rate at which a tape counter should work, so notes made on a recording applied to one type of machine only – not a desirable state of affairs.

The Beocord 8000 was different because it was capable of showing the tape position in minutes and seconds from an absolute reference of zero at the beginning. Previous systems that yielded this level of accuracy had needed a time code to be recorded on the tape when a recording was made but the Beocord 8000 did not need this. Instead, the recorder was able to measure the thickness of the tape (a value that is not constant across the various brands and lengths) by spooling through a measured amount at the start and counting how many turns the take-up spool made as a result. From this figure, any position on the tape could be calculated from simply knowing how many times the take-up spool had turned. The measurement procedure would have been tedious to complete manually but under microprocessor control it all occurred quickly and automatically following the press of one key. Quite a bit of winding back and forth was needed to complete the procedure, so the Beocord 8000 was given a super-fast winding system that allowed the process to be completed in a reasonable amount of time.

Other manufacturers came up with similar real-time tape counters in the following years but few (if any) of these were as sophisticated as B&O's, as most could not be used during fast winding.

To complete the system, the Beogram 8000 turntable was made available. It was desirable to continue the theme of microprocessor control across the system but what could such a complex device do in a simple record player? Quite a lot as things turned out. The tangential tracking system of the previous high-end Beograms was retained but, instead of the arm position being measured with mechanical switches, the microprocessor logged how far it had progressed by counting the revolutions of the lead screw that drove it. The cueing of the arm was also carried out under microprocessor control, with the previous air cylinder damper being replaced by a precision-moving coil actuator, whose position and rate of movement was electronically governed.

The really interesting part of the Beogram 8000 was the motor. Instead of a conventional small motor and drive belt, the Beogram 8000 came fitted with a system known as 'tangential drive'. In simple terms, this took the form of a linear motor wrapped into a circle, so that instead of providing a linear movement, it caused the platter to rotate. Direct drive systems such as this had been a popular feature of top-line turntables for a few years before the Beogram 8000 appeared, but these tended to work in a different way that relied on a more conventional type of motor that could run slowly enough to drive the turntable directly. These other systems were not always successful, as the motors turned in a series of jerks rather than with a smooth action, a problem that B&O's tangential drive did not suffer from at all.

The microprocessor was used to regulate the turntable speed by controlling the power of the tangential drive system. Doing this made the turntable speed perfectly steady, as it was locked to the microprocessor's quartz crystal – the same type of component that made the new digital watches of the period so accurate. The user could alter the speed in 0.05 RPM increments and note the setting from the four-digit LED readout.

ABOVE: With the platter removed the tangential direct drive motor can be seen. One of the driving coils is visible at the bottom left.

BELOW: Removing more parts reveals both driving coils, along with the centre hub and the optical speed sensor.

Along with a lot of new technology, the Beolab 8000 system brought with it a new style for B&O. In the same way that the earlier models had made visual references to the slide rule to suggest precision and accuracy, so the new generation were made to look like computers, the most familiar object of their period that suggested accuracy and precision. The appearance of the Beolab 8000 was both brutal and massive and so fitted in perfectly with the mood of the period.

A complete and functioning Beolab 8000 system is, of course, a prize piece for any B&O collector who has the space to house it. Don't underestimate the room it takes up – the components are very large and difficult to lay out if you don't have the matching SC80 cabinet or SM80 shelving unit, both of which are substantial pieces of furniture in themselves. Each piece of the system is fearsomely complex and unlikely to work properly at the time of purchase. If you are determined to add any 8000 series component to your collection, your time is better spent finding someone who can overhaul it for you, rather than trying to find a perfectly working example. These are precision units that need to be in tip-top condition to give their glorious best. Simply being able to light up and produce a sound of some sort is not enough if you are remotely serious about quality.

New TV Sets

The microprocessor made its entrance into the Beovision TV range in a more discreet manner than it had in the high-fidelity audio line. 1980 would see a new series of models introduced, based around the new Philips 30AX colour tube system. The 30AX represented the final perfection of colour tube design and at last required no adjustments at all beyond basic setting up in the factory. It was also a very slim tube that fitted perfectly into B&O's slender cabinets. The new 30AX models took their styling from a not especially popular range from the previous year (the Beovision 3000, 3300, 3800 and 3900) and would remain in the catalogue until 1987. The initial model was known as the Beovision 8800. This was a 26in table model fitted with infra-red remote control. A simple microprocessor was used to decode the remote control instructions and to operate the basic functions of the set. The Beovision 8800 could also be fitted with a teletext decoder. Teletext was a new information service through which the viewer could read 'pages' of news and information that were selected using the television remote control unit. Although basic in appearance by modern standards, teletext was the first digital information and entertainment format to be aimed at the home user.

LEFT: The curves of the cabinet were wrapped tightly around the slim 30AX picture tube.

BELOW: At a time when some sets still had unsightly cardboard backs, the Beovision 8800 was carefully styled and immaculately finished from all angles.

The Beovision 8800 was the first of a new generation of sets that offered fine performance along with exceptional reliability.
Availability ●●●●●
Complexity ●●

The remote control unit that was supplied with the Beovision 8800 was known as the Beovision Video Terminal. This new unit was a multifunction design that could operate the television, teletext and a video cassette recorder. The slender and distinctive casing was pressure die-cast in metal, which gave it a high-quality feel and a durable finish. This method of construction, along with the basic dimension and scale of the design, would remain a feature of B&O remote controls right up to the present day.

The Beovision Video Terminal was the work of designer David Lewis. He had been responsible, along with Henning Moldenhawer, for some of the more creative Beovision designs of the 1970s but would now become a far more visible creative force at B&O. By the 1990s his designs dominated the range, characterized by their use of gentle curves and moulded plastic, which were so clearly at odds with Jacob Jensen's sleek, slim and ruler-straight styling cues executed in aluminium.

The Beovision 8800 was followed by a whole range of models based around a common chassis (known in some circles as the 33XX) and tube. The buyer could eventually choose between pushbutton, simple remote control and de luxe remote control models, all in a choice of 20, 22 and 26in screen sizes. The numbering of the various models was quite logical: 5000 series models had 20in screens; 7000 series models had 22in screens; and 8000 series models had 26in screens. The other numbers indicated the specification: top-line sets were denoted by having the first and second digits the same (e.g. 8800, 7700). The 20in 5500 was a particularly fine set – B&O were alone in the UK market in using a 20in version of the 30AX tube (rivals used cheaper tubes based on scaled-up portable TV technology for their 20in sets) and were rewarded with a set that produced one of the best pictures ever seen on a domestic television of any period. This series of sets was also exceptionally efficient and reliable. At the time of writing, some examples are

LEFT: The neat and compact layout of the keys set the standard for all future remote controls.

LEFT: The original Beovision video terminal remote control.
Availability ●●●
Complexity ●

RIGHT: The two halves of the casing are die-cast. The strip of clear plastic between them is specially coloured to let only the infra-red signals pass through.

over 25 years old and still going strong – not as cared-for collectables but as everyday sets, a feat that is unlikely to be repeated by any other design.

To complete the range, a cabinet model known as the Beovision 9000 was launched to replace the Beovision 6002. In some countries (notably Germany), two stereo models were launched. These had 22 and 26in screens and were given the numbers 7800 and 8900. They were not marketed in the UK, as no stereo TV material was available at the time.

If you would like an older Beovision set, then these models are the ones to go for. They are reliable enough for regular use and, when in good order, give results that put the current generation of models to shame. The key to their longevity lies in the exceptionally low power consumption, famously lower than a light bulb under normal conditions. Make sure the remote control unit is included and in good condition, and also try to get one with a stand, even if you don't need it right away it makes for flexible placement options later on.

Video Cassette Recorders

The video cassette recorder was the big new product of the early 1980s. Philips had launched the first successful domestic models as early as 1974 but it had taken until the start of the 1980s for the machines to be sufficiently cheap and reliable to be a realistic proposition on the mass market.

B&O had previously made an abortive entry into video recording in the late 1960s by housing a Sony open-reel monochrome recorder in their own wooden cabinet. The model was briefly marketed as the Beocord 4000 but it was not really practical: it could not record TV programmes unless it was connected to a specially adapted TV set and the camera that could be bought with it was bulky and awkward to use, attached as it was by an umbilical cable to the heavy and bulky recorder. The special ½in tape was wound onto open-reel spools and was tricky to thread properly around the exposed video heads, which were easily damaged. It was not until the Philips N1500 video cassette recorder (VCR) arrived that the key features of cassette loading, colour compatibility, a built-in TV tuner and an automatic timer, were all brought together in one domestically acceptable package.

As the technology rapidly advanced, JVC and Sony of Japan came up with their competing VHS and Beta machines, which would, in the following years, slog out a protracted 'format war' in the marketplace. Europe's latest format was the Philips Video 2000 system (V2000 for short). This was the format that B&O chose to back as it entered the video cassette recorder market. Although this can be viewed as a mistake in hindsight, at the time it was a perfectly logical choice, V2000 gave considerably better results than either the VHS or the Beta machines of the day, due in part to the advanced 'dynamic track following' system derived from broadcast practice. This allowed far more information to be packed onto the tape, so much so that the cassettes could be made double-sided just like those used for audio. Dynamic track following also meant that the confusing manual tracking control that was a feature of all early VHS and Beta models was not necessary.

The original Philips V2000 machine (named the VR2020) was so beautifully styled and constructed that it could have been put into the B&O catalogue without alteration. Philips did license the machine to other European brands but the terms were strict – only

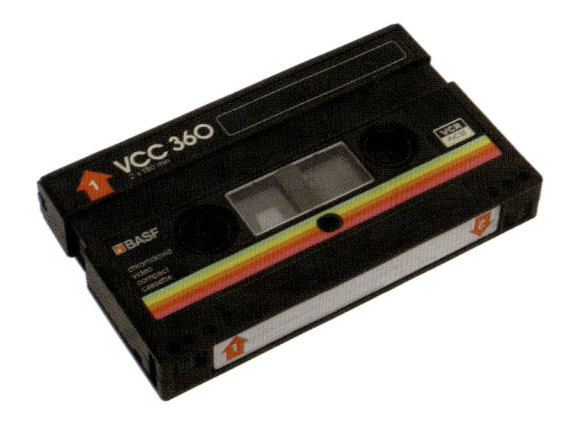

A Video 2000 cassette. Despite its advanced technology, the system was rejected by the market and was an early casualty of the video format wars.

The Beocord 8800V. Note the real rosewood inlays on the sides of the cabinet – you didn't get those on a Japanese machine!
Availability ●●
Complexity ●●●●

the name could be changed, the basic appearance could not. An exception was made for B&O who, in effect, constructed an extra 'outer cabinet' over the VR2020 machine that made it look even more sleek and futuristic. Multiple LED digital displays gleamed out from behind smoked black panels and tiny keys (rather than the big mechanical levers of early Japanese models) operated every function. The programmable timer, the butt of so many bad jokes of the time, was made exceptionally easy to use as it led the user through the programming sequence by a series of illuminated legends. Data entry was made on a calculator-style keypad, rather than by confusing 'up/down' buttons. The Philips VR2020 could be fitted with remote control as an option but B&O's version (known as the Beocord 8800V) included an infra-red remote control system that worked with the Beovision Video Terminal remote control as standard. B&O claimed that their version had improved audio characteristics too, but it is not clear how this was achieved, if indeed it was at all. To complete the B&O treatment, the Beocord 8800V came complete with real wood inlays on the side panels, in either teak or rosewood to match the television it was to be used with.

The model number gave the clue that the Beovision 8800 television set would be the ideal partner for the new recorder and although any of the others could be used, the massive size of the Beocord 8800V meant that the system looked odd if a smaller set was specified.

Philips' restrictive licensing policies led to the unfortunate downfall of the V2000 system. The Japanese manufacturers offered better deals to the European setmakers and distributors, and as a result V2000 would become the first casualty of the much needed rationalization of the VCR market. The very limited availability of V2000 tapes in rental shops would be the format's final undoing; by the mid-1980s, it was all over, despite the millions invested in the superior technology. During this period, B&O had come up with the Beocord 8802V, a model similar to the 8800V but fitted with digital dynamic track following that allowed for 'picture search' and 'still picture' modes. Their final V2000 offering, the Beocord VCR 60, was not so widely distributed and was a compact final-generation Philips model that had received only the lightest of restyles.

False Economies

Until the start of the 1980s, all of B&O's major audio products had been produced in the company's own factories in Denmark. However, in the outside world, the market for high fidelity and music systems, in which B&O competed, had been almost completely taken over by the Japanese marques, a process helped by the abandonment of these sectors by the major European brands. Some had withdrawn completely, whereas others, such as Philips, started to have key models in their ranges made for them by third-party companies, using surplus capacity in Japanese factories. Against this backdrop it was perhaps inevitable that B&O would also eventually have some Japanese-produced equipment in the entry-level positions in their range.

The first model to be imported from Japan was the Beocenter 2000 of 1981. This music centre had a three-band radio complete with four FM pre-tuned programmes, a Dolby cassette deck, a record player fitted with a B&O pickup and a 25W stereo amplifier. Included in the package was a pair of Beovox S 30 loudspeakers, a small model taken from the Uniphase range. This system replaced the very successful Danish-built Beocenter 4600, although the cassette-based Beocenter 2600 remained in the range for one more year. While it was certainly the case that the better Japanese brands were now turning out some beautifully made and beautifully styled equipment, the Beocenter 2000 was poorly constructed and performed indifferently when compared to B&O's own designs. Other than the turntable and the loudspeakers, it was a very ordinary offering indeed, and although cheap for a B&O (which in truth it was in name only), it was expensive compared to what else was available. The control surfaces were made from low-grade matt black plastic and, on the whole, had a cheap, crude action; an indignity, indeed, but worse was to come. The cabinet was trimmed in 'simulated wood', a polite term for grained brown plastic, something that had never before been seen on any B&O product. A music cabinet stand, craftsman-made in Denmark and veneered in real teak or rosewood, was offered to match the Beocenter 2000, but even this could not enliven the drab appearance of the set itself.

One point of interest could be found in the amplifier. It was of a design that used hybrid chip power devices – large, integrated circuits that combined all the transistors required to form the complete power amplifier channel into a single package. While in this case this was certainly done in the interests of economy, this technology would eventually become widespread in mid-range audio and, at the time of writing, is used extensively across the B&O range.

In order to remain saleable, the Beocenter 2000 was updated during the following year to be compatible with metal tape. In truth, metal tapes would have been wasted in such a basic recorder, but at the time, metal compatibility was on every buyer's checklist, so the feature had to be added. The new model was known as the Beocenter 2002.

The Beocenter 2002 was made in Japan for B&O and lacked the quality of finish that the Danish-built models enjoyed. Note the 'wood effect' plastic trim.
Availability ●●●
Complexity ●●

Unfortunately the Beocenter 2000/2002 was not B&O's only attempt to produce an entry-level model that could be made in Japan. The next offering came in 1984 and was known as the Beocenter 2200. Much the same formula was used, with a radio, tape deck and record player packaged together in a one-piece unit. The record player was of similar construction to the Beogram 1800 and the cassette deck used the same mechanism as the Beocord 2000; but, to pay for these refinements, the quality of the supplied Beovox X25 was drastically reduced, with cheap imported drive units, one of which was a rather poor cone tweeter. By this time, wood was out of fashion, so the Beocenter 2200 and its loudspeakers were trimmed in matt grey plastic. Although not a quality finish, this was at least a colour match for some of the more expensive audio products and so did not stand out too much. Again, indifferent sound quality and a poor tactile experience characterized the Beocenter 2200, even if the design of the cabinet was more attractive (from a distance at least) than its predecessor. The user interface looked suitably high-tech with its row of tiny buttons and concealed illuminated radio dial. Remote control could have been a possibility but in practice was never offered; this was a budget product after all.

For the truly poverty stricken, an even cheaper variant was offered in 1986, the Beocenter 2100. This was effectively a Beocenter 2200 without the turntable, and as the turntable and pickup were the only parts that B&O had had a hand in making, this model effectively wrote them out of the picture completely. The public were clearly not fooled, as this model was a slow seller and is seldom encountered today.

Along with the Beocenter 2100, a genuinely interesting model arrived from Japan. The similarly styled

LEFT: The Beocenter 2200 looked more like a proper B&O product than the previous Beocenter 2002 but the quality of materials and construction were still poor compared to the Danish-built models.
Availability ●●●
Complexity ●●

RIGHT: A single wide lid covered the turntable, cassette deck and minor amplifier controls.

ABOVE: The Beocenter 2200 could be considered a styling success, if not a technical one.

LEFT: The turntable and cassette mechanisms were shared with some of the more expensive B&O models.

Beocenter 4000 included a second tape deck and, at a time when dual cassettes were highly fashionable and very much in demand, this seemed like just the thing. At last it seemed that some genuine thought and design flair, along with a slight loosening of the purse strings, had been applied to the Japanese range, as not only was the Beocenter 4000 B&O's only dual cassette machine, it was also the first to offer auto reverse (on one deck only) and their first combination unit to feature the HX Pro recording system. Every attempt was made to accommodate the cassette user, including various sequenced automatic playback modes and a microphone amplifier with its own fader. Sadly the Beocenter 4000 was a better tape deck than it was a complete system, and, hamstrung by the same lacklustre tuner and amplifier that the Beocenter 2200 used, it was still not really up to the B&O standard. However, of all the Japanese music centres it is the most interesting.

In general, unless you simply must have every model, give the Japanese music centres a wide berth. A collector's resources are best employed elsewhere and for equipment of this age, 'the genuine article' costs no more. The original hybrid chips for the amplifiers have not been easily obtainable for some years, so failure (which is far from uncommon) would require complex or costly modifications to put right.

The final Japanese audio product of the 1980s was on an altogether smaller scale. B&O had vacated the portable audio field in 1980 when the Danish-built Beolit 707 transistor radio was dropped from the range. Prices in general had come down a lot and the sector was now fiercely competitive, leaving no room for expensive products like the Beolit.

The most popular type of portable radio in the early 1980s was the radio cassette recorder. The compact size, convenient operation and high quality of the cassette mechanisms of the day made them the ideal partners for the new designs of FM radio that could produce good reception almost anywhere, even when used only with their small rod antennas.

Despite their interests in cassette technology and their close working relationship with Philips, it was looking unlikely that B&O would produce a radio cassette recorder. In the meantime, Sony had proved that it was possible to sell quality models at a premium price, and Hitachi had produced at least one model with many B&O styling cues and details; but for year after year the Beolit range was not revived.

However, the 1987 main catalogue included an unlabelled photograph of an attractive portable stereo radio cassette recorder standing on an outside table. Dealers faced many enquiries about what this was and soon it was to be revealed as the Beosystem 10. No doubt in an attempt to sound modern, the Beolit name had been dropped, but the Beosystem 10 was clearly a continuation of the line. There was one key difference between the Beosystem 10 and the earlier Beolits, for the Beolits had always been designed along the same principles that were common to the

LEFT: **The Beolit 707 was the final model in the series. It can be recognized by its black anodized metalwork. The FM-only Beolit 505 model has a similar appearance.**
Availability ●●●●
Complexity ●

RIGHT: **A close-up view of the new finish.**

rest of the range and were produced to the same standards in the same factories. The Beosystem 10, on the other hand, was a fairly ordinary Japanese stereo radio cassette recorder dressed up in some very smart clothes. The loudspeakers may have looked like miniature CX series Beovox models but in fact they were nothing special, just single 'full range' units mounted in the cabinet with no special considerations (e.g. internal enclosures, reflex loading) made towards improved reproduction. The cassette deck was also a disappointment, having heavy mechanical controls and lacking Dolby noise reduction; the first B&O cassette machine to do so since the Beocenter 1600 of 1976. On the plus side, the radio included B&O's trademark pre-tuned stations, the amplifier was reasonably powerful and the set could be used with a second external tape deck or a record player with ease. The power sources were a choice of mains electricity, eight large flashlight batteries or a 12V DC supply from a car or boat. These details did not make the Beosystem 10 a Beolit, however, but it is of interest to note that the design of parts of the battery holder is very similar to those of the later Beolits, indicating that B&O did at least have some input into the design.

The Beosystem 10 remained in the range until 1990 and during this time it was the cheapest single item in the range that could be used on its own, without the purchase of further equipment. Despite this, it was still expensive and some Japanese branded equipment (notably the offerings from Sharp and Aiwa) offered better performance and more features for less money in some very attractive packages. Philips and Sony had also successfully integrated compact disc players into radio cassette recorders at an almost affordable price, by this stage, and it was these that drew the interest of the wealthy portable buyer.

The Beosystem 10 has proved just as popular as the previous Beolit series with collectors, so there are few real bargains to be had. If considering one, ensure that it is complete, as the chance of finding any missing parts is slim. When contemplating the purchase of an expensive example, take time to think whether it represents good value, given the minimal Danish content.

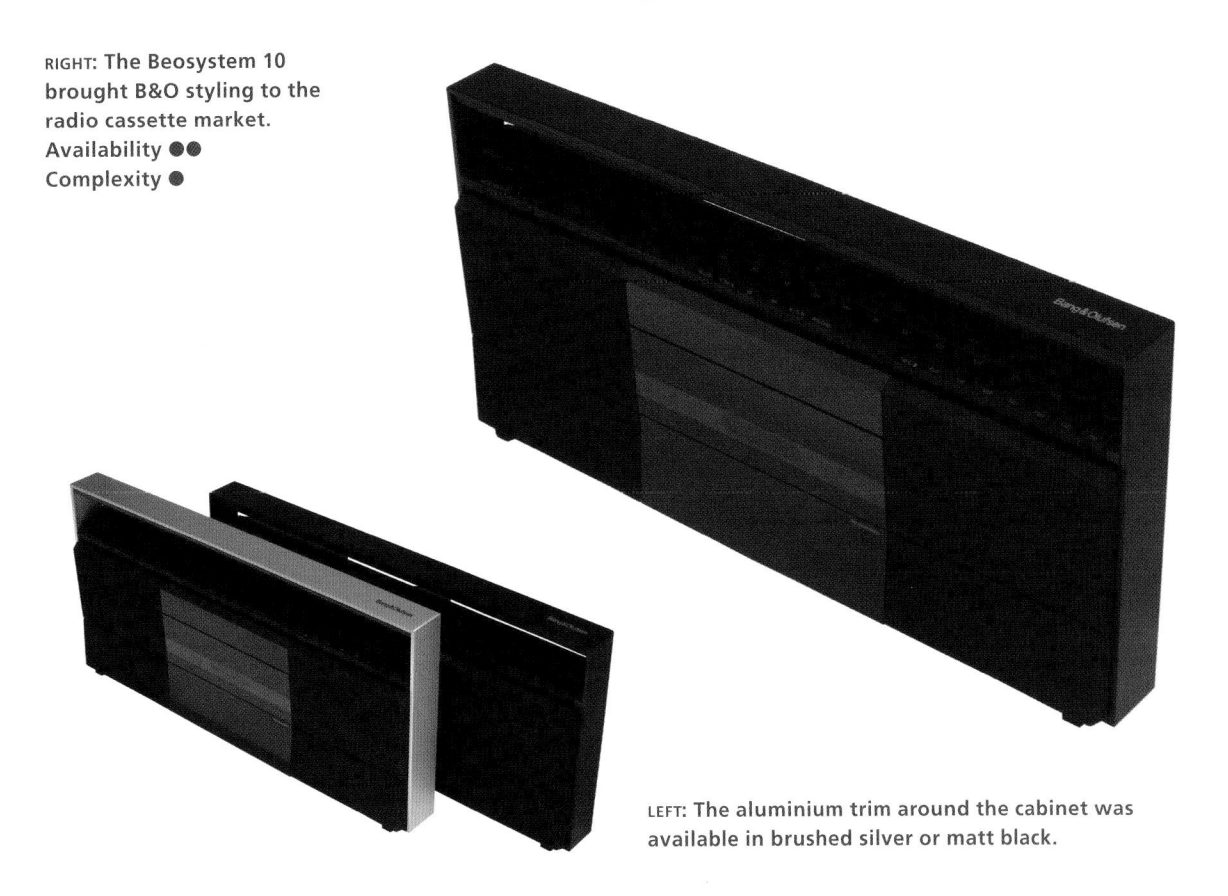

RIGHT: The Beosystem 10 brought B&O styling to the radio cassette market.
Availability ●●
Complexity ●

LEFT: The aluminium trim around the cabinet was available in brushed silver or matt black.

The end of the Beosystem 10 saw the Japanese influence on the B&O range retreat back to its natural base of providing technology and manufacturing skill for video cassette recorders and mechanisms for audio cassette decks. Other than in these two areas, and the supply of minor electronic components, Japanese products would not return during the period covered by this book.

Smaller Beomaster Models

Impressive though the Beomaster 8000 was, its size and cost made it not practical for everyone. Because of this, it was seen as important to maintain a range of smaller Beomaster models, along with suitable source equipment to use them with, in the range. The Beomaster 2400-2 was the key model, partnered with the choice of turntables and either the Beocord 1900 or the later (and much improved) Beocord 2400 model.

The Beomaster 1500 was discontinued in 1980 and replaced by the Beomaster 1700. This AM/FM set had a striking appearance. It was slim, sleek and topped by a glass plate onto which the tuning scales and those for the volume, tone and other controls were printed. Aluminium keys along the bottom edge engaged each function, which could then be adjusted by rolling rubber tracks located at the edges of the glass plate. The gears and clutches that allowed the control system to operate recalled the complexities of the Beomaster 6000 4channel, although the removal of the motor and its attendant control system meant that, in practice, the system used in the Beomaster 1700 was considerably simpler. The Beocord 2400 was re-packaged in a new cabinet to match the Beomaster 1700 and was of course known as the Beocord 1700.

1980 TO 1989: THE MICROPROCESSOR AGE

An all-black version with an FM-only radio was also made, known as the Beomaster 1600. The matching cassette deck for this model was called the Beocord 1600 and both are quite unusual. Years of contact with greasy fingers have, in some cases, caused the rubber tracks to weaken and disintegrate – collectors should be careful when inspecting these models, as original replacements are no longer available.

1984 saw the arrival of a new small Beomaster design that continued the Beomaster 1900 line but included some new technology, such as Datalink control for the external sources and a hybrid power

RIGHT: The Beomaster 2000 and Beocord 2000 together showed that at last the receiver and the cassette deck were a perfect match.
Availability ●●●
Complexity ●

As beautiful as ever, the Beomaster 2000 continued the 1900 line.
Availability ●●●
Complexity ●

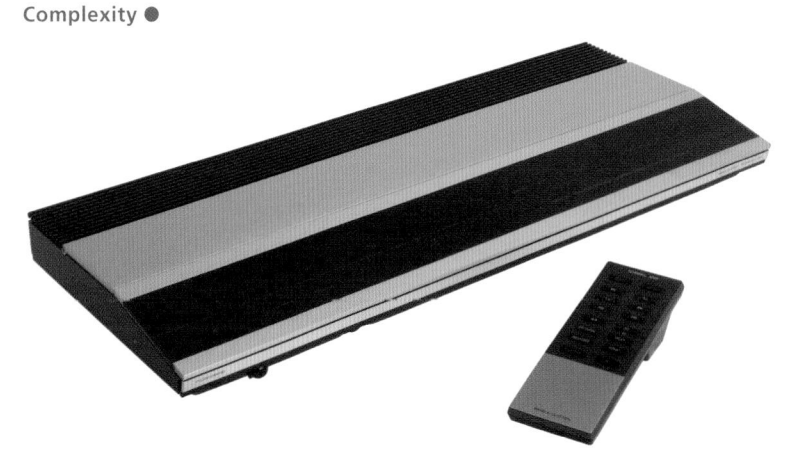

The remote-controlled Beomaster 3000 replaced the earlier 2400 model.
Availability ●●
Complexity ●

ABOVE: The infra-red remote control unit for the Beomaster 3000 was styled similarly to that provided with the Beolab 8000.

1980 TO 1989: THE MICROPROCESSOR AGE   **105**

LEFT: **The Beocord 2000 was the first B&O cassette deck to use touch-sensitive controls.**
Availability ●●●
Complexity ●●●

RIGHT: **With the lid closed, the uncluttered lines meant that it was not immediately obvious that the Beocord 2000 was a cassette recorder at all.**

amplifier chip. Two models were initially offered: the Beomaster 2000, without remote control; and the Beomaster 3000, which, for the first time in the smaller Beomaster range, included an infra-red remote control unit.

These advanced models, which headed up the Beosystem 2000 and 3000 ranges, took some time to prepare, and so, for 1983 only, the stop-gap model Beomaster 2300 was offered. The Beomaster 2300 was essentially a Beomaster 2400 cabinet with Beomaster 1900 internals and, therefore, lacked a remote control system.

The Beomaster 2000 and 3000 were thoroughly modern. Their display panels finally dispensed with incandescent lamps and relied instead on LEDs only. Their cabinets were trimmed in dark grey plastic, instead of real wood, which, although regrettable, now reflected perfectly the fashions of the day. The matching Beocord 2000 (applicable to both models) for the first time featured electronic touch controls instead of mechanical keys and was slim enough to be not only identical in profile to the Beomaster but identical in height too.

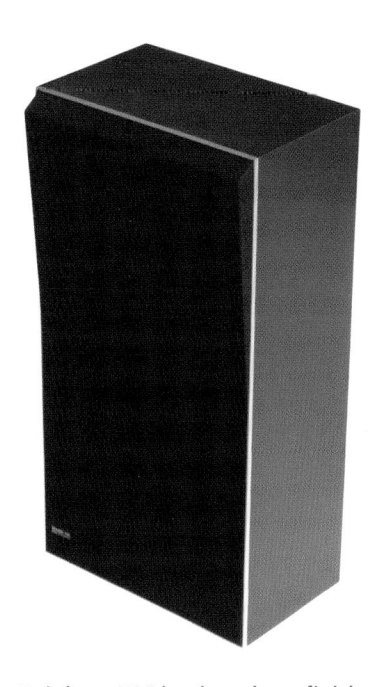

Beovox Uniphase X35 loudspeakers, finished in matching metallic grey, were one option to complete the new 2000 and 3000 systems.
Availability ●●
Complexity ●

ABOVE: **The Beogram 3300, the turntable for the Beosystem 3300.**
Availability ●●●
Complexity ●●

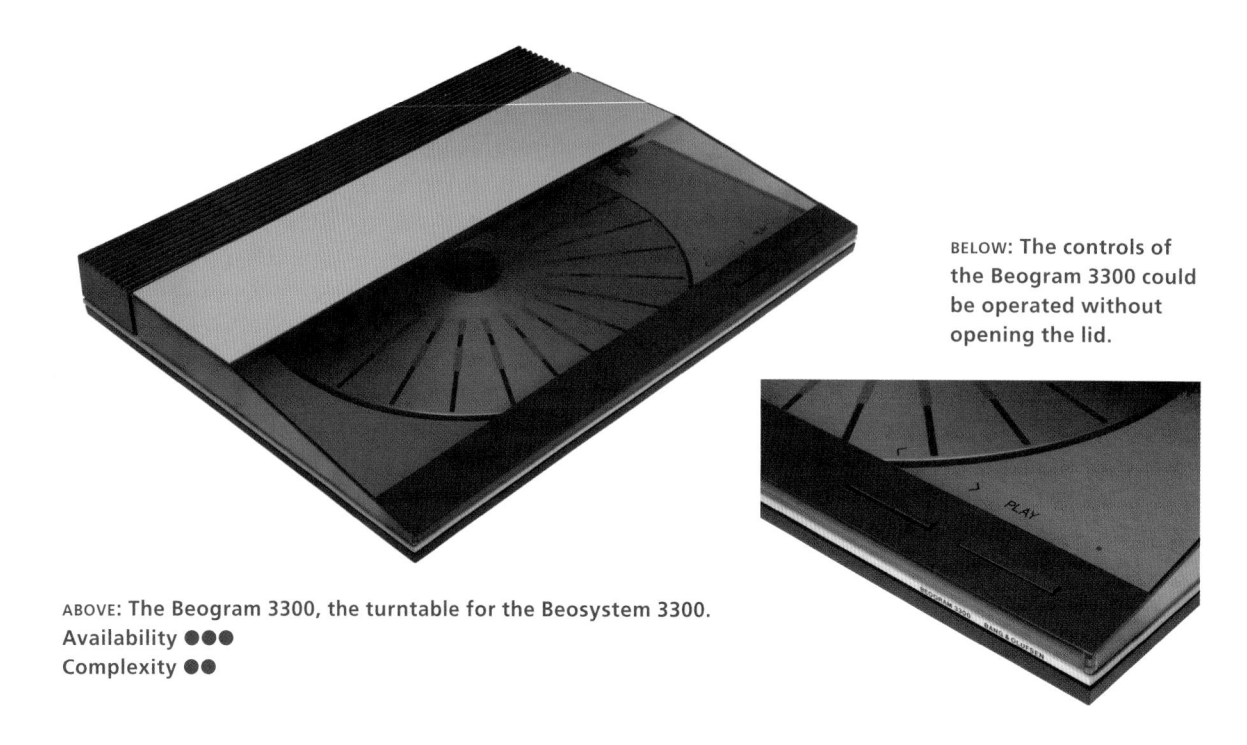

BELOW: **The controls of the Beogram 3300 could be operated without opening the lid.**

Two Beogram models were produced to complete these systems. The Beogram 2000 was a radial model based on the Beogram 1800 but with Datalink and a repeat function. The Beogram 3000 was part of a new range of compact linear tracking models and is easily recognized by its angled lid and black platter.

This pair of compact systems remained in the range until 1988 when both were replaced by the Beosystem 3300. This new series was a development of the Beosystem 3000 and, importantly, introduced a specially designed compact disc player as part of the system. The light grey finish with bold text and orange lighting are not to all tastes, but again were the height of fashion at the time.

Developing the Beolab 8000

One of the functions of a product range as striking and as uncompromisingly 'hi-tech' as the Beolab 8000 system was as a showpiece of B&O's advanced technology. Even if they could not afford to buy it, potential customers would be impressed and so look favourably on the more affordable offerings at the more sensible end of the range. To fulfil this role, the Beosystem 8000 had to be kept up-to-date, and so as technology and technique advanced, so the various units were improved.

The first part to be updated was the Beocord 8000 cassette deck. After only a year this was replaced by the Beocord 8002, an almost identical new model with an important new feature. The Beocord 8002 included a recording system called Dolby HX Pro. As the name suggests, this was developed in conjunction with Dolby Laboratories but it had nothing to do with the Dolby noise reduction system. 'HX' stood for 'headroom expansion' and was a system that carefully adjusted the recording characteristics of the machine to match the content of the programme that was being recorded. For quiet passages, the conditions were optimized for low noise, but for high energy music with lots of treble, they were instantly and automatically altered to give an increased overload ceiling. This new technique improved the recording quality that could be achieved, even with quite cheap ferric tapes. It was later adopted under license by some famous Japanese manufacturers. Part of the licensing conditions specified that B&O were to be credited as the originators of the system somewhere on each machine, so their name can be found in some quite unexpected places.

As HX Pro worked during recording only, the advantages it gave could be enjoyed on any tape recorder that was used to play back an HX Pro recorded tape. B&O included it on many of their better models right up until the end of their involvement with the cassette system, whilst Sony still have an HX Pro recorder in their range, at the time of writing.

The Beocord 8002 could also recognize IEC standard metal tapes (type IV) automatically; previously a manual setting had to be made if these were to be recorded on.

The next development of the Beocord 8000 series marked another great step forward in cassette recorder design. The Beocord 9000 of 1982 included a computer-controlled calibration (CCC) system that matched the recorder to the particular tape that was loaded, guaranteeing a perfect recording every time. The process worked by recording a series of test tones onto the tape and then measuring the resulting playback signal. Bias, equalization, recording current and meter calibration were then optimized automatically. To fit this complex procedure into a reasonably short length of time, separate record and playback heads were fitted, although, despite this, 'off the tape' monitoring was not possible. CCC was undoubtedly a technical triumph but in truth it arrived a little too late. The IEC had taken steps to standardize tape formulations and so by the time the Beocord 9000 was ready, the problem of the large spread of tape characteristics, which had dogged the industry throughout the 1970s, had largely been solved. CCC did, however, compensate up to a point for wear in the tape heads and drift in the electronics of the machine, so the need for regular maintenance was reduced. It also allowed good recordings to be made despite 'batch variations' that still occurred with some makes of tape. Because of the wide range and versatility of CCC, the Beocord 9000 was also the only Beocord cassette recorder that could make correct recordings on 'ferrochrome' tape (IEC type III) of the type popularized by Sony.

LEFT: Comprehensive controls were hidden under an aluminium lid.

BELOW: At a glance, the Beocord 9000 looked just like the other models in the series but the CCC system made it very much more complicated and expensive.
Availability ●●
Complexity ●●●●●

108 1980 TO 1989: THE MICROPROCESSOR AGE

As well as CCC, the Beocord 9000 also included Dolby C-type noise reduction. This offered an improved performance over the standard B type of around 10dB, a very worthwhile improvement. In addition, the real-time tape counter was redesigned and enhanced, so that it could indicate how much time remained on a cassette during recording. This feature was combined with a two-speed winding function that reduced the chance of tape breakage at the end of each side.

Each Beocord 9000 was originally supplied with a special tape and screwdriver, so that the owner could adjust the head azimuth back to the factory setting should the adjustment deteriorate or if the heads had to be adjusted to play a recording correctly that had been made on a poorly set-up recorder. The cassette used was a TDK MA-R type, an expensive precision-built item using a die-cast frame as its basis.

A TDK MA-R (metal-alloy reference) cassette, the height of precision in the cassette system.

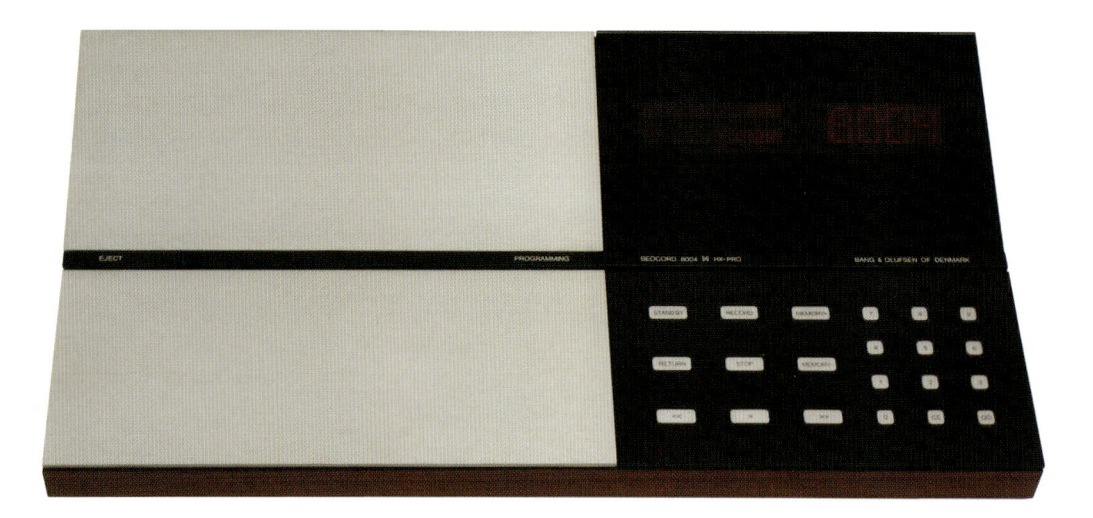

ABOVE: **Thanks to the standardization of recording tapes that was put in place in the early 1980s, the Beocord 8004 could produce recordings of outstanding quality without the need for the complex CCC system.**
Availability ●●●
Complexity ●●●

RIGHT: **The addition of Dolby C and automatic metal tape recognition caused a minor rearrangement of the control panel.**

In the following year, the Beocord 8002 was replaced by the final version in this series, the Beocord 8004. The Beocord 8004 included the same Dolby C noise-reduction system that the Beocord 9000 had offered a year earlier, but it was implemented using two powerful Dolby processor chips from Hitachi, instead of the rat's nest of separate components that the 9000 had used. The Beocord 8004 was a very fine recorder indeed and when set up carefully was capable of top-class results. Around the same time, the Beocord 9000 was redesigned to use the new simpler (and better) Dolby circuit, and numerous minor circuit modifications were made to centre the characteristics of the CCC system around the new types of IEC standard tapes that the major manufacturers (e.g. TDK) were now producing.

Of all the models in the Beocord 8000 series, the Beocord 8004 is the one of choice. Carefully set up it is capable of outstanding results and from the start was designed to match IEC standard cassettes perfectly. The Beocord 9000 is of course intriguing but it is also a complex beast with many specialized parts. The CCC system helps a lot to compensate for wear but careful setting up and calibration of the base adjustments is still occasionally needed to keep the machine on top form. All models in this series suffer from missing elements in their LED readouts; this is annoying and spoils the look of the machine as a whole. Complete display units are no longer available but with care they can be opened up, the defective LED chips removed and replaced with individual miniature surface mount components. Done with care, this repair is undetectable from the outside, but it requires skill, dexterity and patience to complete.

In 1984 the Beogram 8000 was updated. The reason for the change was that a new range of miniature pickup cartridges had been introduced to replace the MMC 20 line that had first been seen fitted to the Beogram 4000 of 1972. The new range had a simple numbering system, the letters MMC were followed by a number from 1 to 5, with 1 being the finest and 5 being the most basic. The difference was mainly to be found in the cut of the diamond stylus and the materials employed for the cantilever. To use this new pickup, the Beogram 8002 (as the new model was known) was fitted with a revised arm. At the same time, a new design of anti-static platter was also fitted and the programming of the microcomputer revised.

The new style of miniature MMC pickups, as first seen in the Beogram 8002.

The choice of Beogram 8000 or 8002 should not nowadays be based on the technical merits of either model; rather the quality of the actual example on offer should be used as a guide. One improvement that was made to the later version was that the sensor disc located beneath the platter was changed in construction from printed acetate to punched metal. The earlier type tends to degrade and cause speed jerks or irregular operation and satisfactory replacements are difficult to find. Check the condition of this key component before buying either model.

The Beomaster 8000 was never officially revised and retained the same numbering throughout. There were some detail changes made, however, the most obvious of which was that the tape monitor function was moved from the TP1 (tape 1) input to the TP2 one.

'The Poor Man's Beolab 8000'

For those who really liked the Beolab 8000 system but found it too large, expensive, complex or perhaps all three, B&O offered two alternatives. Both were designed along the basis of offering a similar level of quality but with fewer features and less power in a more compact size and both would find their own market niches.

The first to appear was the Beocenter 7000. This combined an amplifier, an AM/FM radio, a cassette recorder and a record player into one reasonably compact unit. Introduced for the 1981 season, the Beocenter 7000 was a machine of exceptional quality. It was also technically advanced and included a computer-controlled user interface with a programmable timer to either turn the set on or to make unattended recordings. Remote control was also standard, using a similar infra-red terminal unit to the one that was supplied with the Beolab 8000 system. Also, shared with the 8000, were the cassette mechanism and numerous design details of the rest of the circuitry. The amplifier of this model was rated at 40W, a healthy figure for a combination unit.

The Beocenter 7000 sold strongly and was soon updated with the arrival of the Beocenter 7002. This model could be used with metal tapes and also included a new lighter design of pickup arm. The Beocenter 7002 was joined in the range by the Beocenter 5000, a stripped-down version that did not include the electronic timer, the computer-controlled tape counter search function or the remote control. The Beocenter 5000 could be recognized by its black metal trim, as all the other versions were finished in natural aluminium.

1983 would see the finest version of this series launched. The Beocenter 7700 looked similar and was constructed along similar lines but included numerous technical and styling improvements over the previous versions. The most noticeable change was the record player lid, which was finished in aluminium to match the cover over the minor controls and the cassette compartment.

RIGHT: **The Beocenter 7700, finest of all music centres.**
Availability ●●●
Complexity ●●●

LEFT: **With the lids open the complete choice of top quality sources can be seen.**

ABOVE: **Digital displays and controls like a calculator keypad looked very futuristic in 1983.**

ABOVE RIGHT: **The small pillar to the right of the pickup is not a rest for the arm but a neatly styled catch for the lid. Pressing the black plastic strip at the front caused the lid to open automatically.**

ABOVE: **This profile view shows just how slim the Beocenter 7700 is.**

The record player lid now opened by itself at the push of a button and had an automatic light inside to illuminate the record; the control panel and display had also been tidied up; and, finally, the old slide volume control was replaced by a set of electronic push-buttons. The record player made use of the same new miniature pickups that the Beogram 8002 had used and, depending on how generous the factory were feeling, sometimes a quite exotic type was fitted, an MMC 3 or even MMC 2 in some cases. The electronic timer was also expanded so that it could now be programmed for two separate events. All these improvements, however, were mere sideshows when compared to the big new change: two-way remote control.

The two-way remote control worked by having both a transmitter and a receiver in both the set and the remote control unit. Using this bi-directional link, it was possible for the set to report back that an instruction had been successfully executed (or not) and for the operating status to be checked at any time.

Two-way remote control made possible and practical a completely new way of using a quality audio system. The multi-room setup expanded the established practice of putting a pair of extension loudspeakers in a second room, into an elaborate arrangement where the system could also be operated from that room as well. A control unit and an infra-red transceiver were mounted in the second room along with an extra pair of loudspeakers. All that was then necessary was to run a special cable back to the main unit and then the system was effectively available in both rooms. Up to three extra rooms could be connected in this original version of the system. Multi-room operation became another B&O hallmark and for years after the launch of the Beocenter 7700, no other manufacturer offered anything similar. The same multi-room equipment, which was intended for the Beocenter 7700, could also be used with the older Beocenter 7000 and 7002 models without modification.

The Beocenter 7700 was augmented by a simpler version, the Beocenter 7007. This lacked the two-way remote control and the aluminium cover over the

ABOVE: The shape of the Master Control Panel 7700 made the unit easy to use without having to pick it up.

LEFT: The infra-red transceiver in the Beocenter is hidden inside these three black lenses.

TOP LEFT: The system in the palm of your hand – the first two-way remote control unit, the Master Control Panel 7700.

TOP RIGHT: The Beocenter 7700 could also be operated with the smaller Beocenter Terminal remote control unit.

BELOW LEFT: Compact, high-quality loudspeakers were often needed for a multi-room system. The Beovox CX 100 proved ideal.

record player (which also came fitted with a more basic pickup) but was otherwise identical to the Beocenter 7700. It was this model that would endure the longest, being deleted from the range in 1987. It was the last B&O main audio product to feature real wood trim. After that, cheaper and less attractive synthetic materials predominated.

All these music centres can be bought for a very reasonable price but at this level don't expect everything to be in perfect working order. The cassette deck will usually be in need of a mechanical overhaul (tricky but it can be done, if care and time are taken) and various other minor functions may not work. Overhauling any model in this series is time and money well spent; they are one of B&O's nicest products.

Another alternative to the Beolab 8000 appeared for inclusion in the 1982 range. Rather than being a combination unit, the Beolab 6000 was, like the 8000, a three-part component system based around a scaled-down Beomaster receiver.

The Beomaster 6000 looked like a slightly smaller Beomaster 8000 but, in reality, it owed more to the Beomaster 4400 in terms of electronic design. The 4400 circuit was re-worked and fitted into the cabinet of the Beocord 8000, to which an extra compartment had been attached at the back to house the mains transformer and the power amplifier. The Beomaster 6000 was rated at 75W per channel, half of what the

The Beomaster 6000 offered a tempting alternative for those who found the Beomaster 8000 too large or too expensive.
Availability ●●●
Complexity ●●●

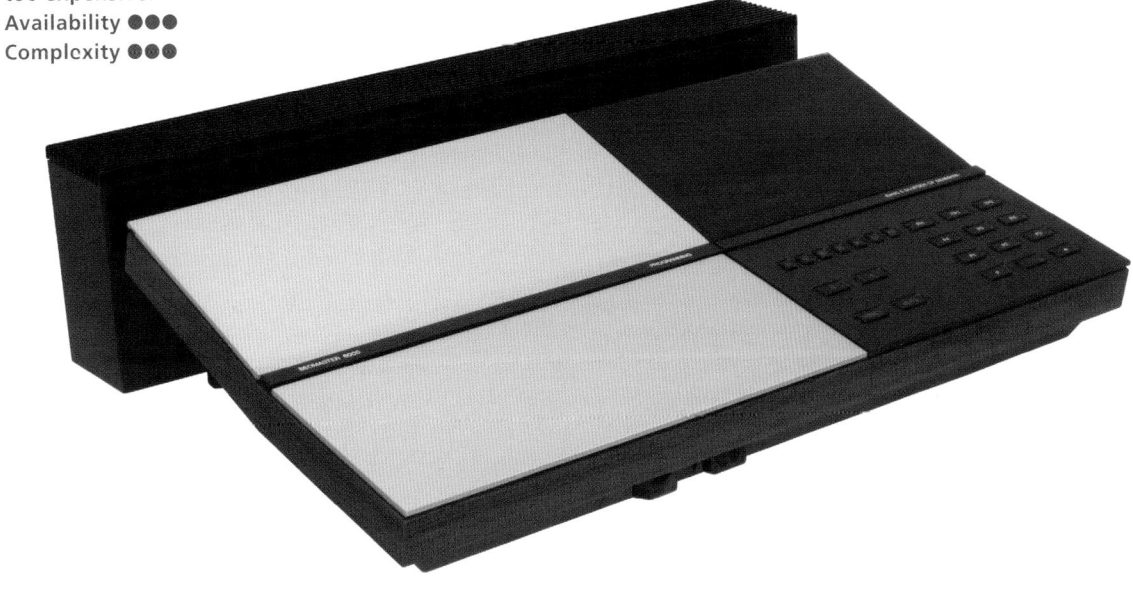

Behind the high-tech fascia, the Beomaster 6000 used technology common to some of the earlier Beomaster models.

Pushbutton switches located in a recess behind the lid allowed two pairs of loudspeakers to be selected.

Beomaster 8000 could produce but roughly twice as much as the Beocenter 7000 series was capable of.

To maintain the family look it was desirable to equip the Beomaster 6000 with electronic controls and a digital display. The Beomaster 4400, however, had done without these things and so adding them retrospectively was not going to be an easy task. The design route chosen was to 'overlay' a digital interface onto the conventional analogue circuitry that had been brought over. This was done, for example, by retaining the traditional voltage controlled tuner (first seen in the Beolit 500 of 1965) but using a digital frequency counter (instead of a traditional tuning dial) as the display. Pre-set tuning was offered by including the familiar bank of miniature tuning wheels, whilst the main tuning was performed by a large, heavily weighted thumb wheel. Powerful automatic frequency control (AFC) and an effective muting circuit were used in combination to give an impression of 'search tuning' as the FM band was scanned, and although not sounding particularly promising in concept, on the whole the system worked really well.

The volume control was another area where digital controls were applied to conventional components. Instead of the complex electronic attenuators that were found in the Beomaster 8000, the Beomaster 6000 used a conventional rotary volume control driven by a small motor. The volume control was buried deep inside the set and all the user saw were the electronic controls and a sliding bar-graph display. The display looked impressive but was, in fact, simply a strip of red transparent plastic that was pulled along by a cord wrapped around the motor shaft. Careful styling disguised this quite well, however. The volume control system was designed to mirror the functions of the electronic arrangement in the Beomaster 8000 and this made it very complicated. The motor could be slaved to a traditional sliding control so that the volume level that the set started on could be set and 'stored'. In addition, seven keys gave (almost) instant access to selected points along the volume scale, evenly spaced from zero to full output. Why anyone would want to instantly summon up so much power was not explained; it was a function that was probably seldom (if ever) used. One electronic function fitted to the Beomaster 6000 that the 8000 did not have was a digital clock and timer. This could be used to set the system to turn on at any time, although if a recording was to be made, the clock in the Beocord would have to be set as well. The Beolab 8000 could also be set to make timer recordings but this relied on the clock in the Beocord alone.

The Beomaster 6000 was not initially fitted with remote control as standard and the kit to convert it was expensive. If the buyer were to opt for it, the Beomaster 6000 was not greatly cheaper than the 8000 and, therefore, did not offer particularly good value for money. Despite this, its smaller, neater size was attractive to some and the audio performance was very good.

To complement the Beomaster 6000, a range of matching source components was introduced. These were based on those offered for the 8000 range but were of a slightly lower specification. The Beocord 6000 was, in simple terms, the same as Beocord 8000, which had by this stage been replaced by the Beocord 8002. The advantages of the Beocord 8002 were brought to the 6000 range with the introduction of the Beocord 6002, again only made available once the Beocord 8004 had been launched for the top-line system.

Two turntables were offered. The first was the Beogram 6000, a modified version of the Beogram 2400. Even though this had been fitted with the Datalink system, the styling match was poor and this model did not prove popular. The second choice was the Beogram 6006. This was more expensive and was very similar to the Beogram 8000, the only real difference being that a pickup of a slightly lower specification was fitted. Once the Beogram 8002 arrived, the turntable for the Beolab 6000 system was revised too. The new model was known as the Beogram 6002 and this looked very similar to the 8002. Inside, however, a great deal of simplification had taken place as the microcomputer and the tangential drive system had been removed and replaced by a simple logic control circuit and a conventional DC motor/belt drive arrangement. This new and simpler model helped to maintain the distinctive difference between the 6000 and the 8000 ranges. It was also offered without

Datalink as a non-committed stand alone player for non-B&O systems in which form it was given the name Beogram TX.

All 6000-series components could be recognized by their black plastic control keys – those of the 8000-series models are finished in natural aluminium. The Beocord 9000 has a mixture of keys finished in both materials to match both systems.

The Beomaster 8000 was last sold in 1984 and from this point onwards, to the deletion of the whole series at the end of the 1986 season, a simplified range was offered. The Beomaster 6000, Beogram 8002 and a choice of either the Beocord 8004 or the Beocord 9000, were all that was available. Even in this reduced form, the Beosystem 6000, as the range was now known, was a very serious high-fidelity setup indeed and, although it was not widely recognized or acknowledged by the British hi-fi press, its performance still far exceeded that of the many finicky, ugly and difficult to set up systems that they routinely recommended at the time.

The Beomaster 6000 is an attractive piece and certainly deserves a place in anyone's collection. If anything, it is slightly sweeter sounding than its big brother, and is certainly easier to house. The power amplifier has proved to be a little fragile in service, so do ensure that both channels still work. Another common failing is that the tuning display fails to work, showing '88.8' or some other incorrect reading, regardless of which station is tuned in. While this does not make the radio unusable, it is annoying. The cause is almost always a defective pre-scaler chip, for which it can be difficult to find a replacement. As not all Beomaster 6000s came with a remote control, it is nice to find one that does – but ensure that you are given the correct one.

A Stack System at Last

Since the start of Unit Audio in the late 1960s, B&O's designs had been in the main long and low, intended for shelf or table placement. This gave them an elegant and sleek look in the catalogues and showrooms but to achieve the same look at home, required a lot of space or some very carefully chosen furniture.

The rest of the hi-fi industry had solved this problem with the stack system. In a stack system, each component (tuner, amplifier, tape recorder) was made to be the same width and depth, so that they could simply be placed one on top of the other. A good idea in theory, but a poorly designed stack system could, in practice, be a real eyesore, covered in buttons, displays and knobs set against a backdrop of mismatched surface finishes. Examples of these design pitfalls proliferated throughout the 1970s and 1980s, but, on the whole, the buying public liked what they saw and bought stack systems in truly massive quantities.

B&O's designers finally relented and produced a stack system. The stunning Beosystem 5000, shown here on its specially designed stand, was the result.
Availability ●●●●●
Complexity ●●●

116   1980 TO 1989: THE MICROPROCESSOR AGE

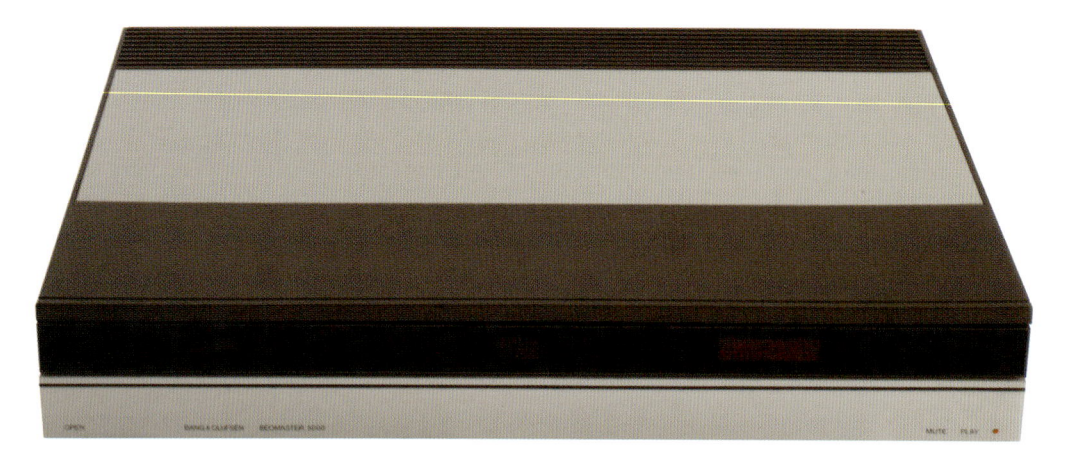

ABOVE: **The Beomaster 5000 was the centrepiece of the Beosystem 5000.**
Availability ●●●●●
Complexity ●●●

RIGHT: The front of the Beomaster 5000 could be opened to gain access to the control keys. This feature was dropped from the later models in the series.

The insides of the Beomaster 5000 were so densely packed that a fan was needed to keep the powerful amplifier cool.

The stack system format was something that B&O had resisted. The nearest they came to producing one was the Beolab 1700 of 1973, which could be stacked up neatly with the matching Beocord 1700 cassette deck on the top, leaving no room for the turntable.

The first complete B&O stack system came in 1984 with the introduction of the Beosystem 5000, which, despite being placed near the top of the range, sold in large numbers. As initially offered, it comprised a receiver, a turntable and a cassette deck, all of identical dimensions and appearance. As ever, the centrepiece of the system was the receiver and this was a work of art.

The Beomaster 5000 was an AM/FM receiver with an output rated at 55W. It was loosely based on the relevant stages of the Beocenter 7700 but with countless changes and improvements.

The Beomaster 5000 bristled with 'firsts', for as well as being the first Beomaster specifically designed for stacking, it was also the first with digital tuning and station memory on both AM and FM, the first to offer full Datalink control over two cassette decks, the first with forced air cooling by an electric fan, the first with a fully programmable multi-event electronic timer and the first to offer electronic tone controls that could be adjusted remotely.

The basic functions of the Beomaster 5000 could be operated by opening the front panel but for full control the Master Control Panel 5000 was needed. This was like a larger and more comprehensive version of the Master Control Panel 7700 that had been included with the Beocenter 7700 the year before. Physically larger and much more comprehensive, the Master Control Panel really did put the whole system in the user's hand. Unlike the Beocenter 7700, there were some functions of the Beosystem 5000 that could only be operated using the Master Control Panel, such as the timer and adjustment of the tone controls.

TOP LEFT: **When launched, the Master Control Panel 5000 was the most advanced remote control unit in the world.**
Availability ●●●●
Complexity ●●●

MIDDLE LEFT: **The Master Control Panel does not look nearly as slim from the side, the bulge on the underside was necessary to accommodate the four D-size cells required to run the unit.**

BOTTOM LEFT: **Opening the hinged cover revealed extra controls for programming the system and a brief operating guide.**

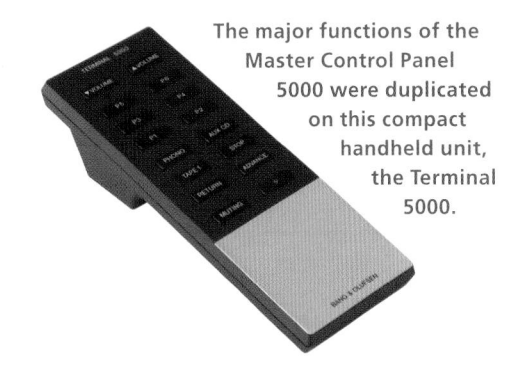

The major functions of the Master Control Panel 5000 were duplicated on this compact handheld unit, the Terminal 5000.

BELOW: **This side view of the Terminal 5000 shows how the battery compartment defined the slope of the keypad.**

The Beocord 5000 cassette deck again showed the design creativity and skill that was present at B&O during this period. The design of all the Beosystem 5000 components required them all to be the same size and for their positioning to be as 'free' as possible. The only way a cassette could be accommodated in the space was to lie it down, but for traditional designs this posed a problem of access, as it is difficult to design a stack system around a top-loading cassette deck. The answer was to allow the whole body of the Beocord 5000 to slide out from the outer cabinet under motor power. This method of construction, though complex and expensive, not only solved all the problems but gave Beosystem 5000 instant high-tech appeal and showed that it was clearly different (and better) than other stack systems. The recorder could be operated with the 'drawer' either open or closed, and recording (or playback) was not interfered with in any way, even if it was opened or closed while the tape was running. Technically the Beocord 5000 owed a lot to the Beocord 8004 but lacked the real time tape counter and used a simpler mechanism similar to that of the Beocord 2000.

The record player for the Beosystem 5000 (naturally given the name 'Beogram 5000') had been seen in the previous year's catalogue in a simplified form known as the Beogram 1800. The Beogram 1800 was not intended for any system in particular, but it was a natural styling match for the Beomaster 2000 that would shortly follow. It used a new mechanism that was very

The control keys of the Beocord 5000 could only be reached with the drawer open.

slim and fast-acting, along with an incredibly delicate-looking slender tone arm tipped with the latest MMC series pickup. To this already desirable machine, the designers added a tinted lid with aluminium inlay and an automatic opening function, an automatic light to illuminate the record, Datalink remote control and new styling to match the other Beosystem 5000 components. Both the Beogram 1800 and the Beogram 5000 were based around a rigid cast-resin chassis and were well-finished and capable of very credible performance. Although it was not apparent at the time, this series of machines would be the last design of radial turntable that B&O would market. The final model was the Beogram RX 2. This was a

Making the Beocord 5000 open like a drawer solved the problem of integrating a high-quality cassette deck into a stack system.
Availability ●●●●●
Complexity ●●

RIGHT: **A light in the lid of the Beogram 5000 helped the user cue up records accurately. With the lid closed, the Beogram 5000 looked just like all the other Beosystem 5000 components.**
Availability ●●●
Complexity ●●

LEFT: **The incredibly slender arm was tipped with an MMC 4 pickup.**

non-dedicated turntable intended for any make of audio system and was sold at a very attractive price, made possible partly because the pickup cartridge was not included.

In its initial form the Beosystem 5000 was clearly just what B&O's customers wanted. Its desirability was further enhanced by the inclusion of full multi-room facilities, just like the Beocenter 7700 offered, but made even more versatile by the technologically advanced Master Control Panel 5000. Placement options were varied but most owners opted to stack the units one on top of the other. As an optional extra, the SM50 stand could be ordered to hold the stacked units, which also offered useful storage space for cassettes and other small accessories. The system received two important developments in its lifetime, the first of which was a new type of turntable. The Beogram 5005, as it was confusingly named, was a compact linear tracking model, whose styling hinted at that of the Beogram 4000 and 8000, two prestige models famous for their quality performance.

The Beogram 5005 used a new mechanism that enabled the construction of a tangential tracking turntable to be considerably simplified. Instead of moving the arm across the record using a precision lead screw, it was pulled along by a drive cord. Even though the second arm was retained to show styling continuity, it was no longer used to sense the record size; instead, the records were weighed by raising them briefly on the centre puck. Tangential drive was also dispensed with; a DC servo motor coupled to the platter by a flat rubber belt was used in its place. The Beogram 5005 and its many derivatives were the last series of turntables to be marketed by B&O, the final version being the Beogram 7000.

The second development was the release of a matching compact disc player to complete the system. The Master Control Panel 5000 was from the start equipped with keys for compact disc track selection and programming, so owners knew that it was only a matter of time before a suitable unit was produced. The Beogram CD 50, as the model was known, is fully described in a later section.

The Beosystem 5000 became an instant classic at the time of its launch. Such was the pent-up demand for a B&O stack system that many thousands were sold, leaving rich pickings for the collector these days. Try to buy a complete set – it can be frustrating trying to build up a system from individual components. Make sure that the Master Control Panel is included with the Beomaster, as many of its functions cannot be used without it. The tops of all these units mark rather easily, so if you wish to present the system in anything other than a stack, inspect them carefully first. The correct order to stack the complete system, starting at the bottom, is: Beocord, Beogram CD, Beomaster, and Beogram. This order ensures that all the controls are always accessible.

From Uniphase to Red Line

The larger models in the Uniphase loudspeaker range received an update in 1981. Three new ranges, two of three-way models and one of large four-way designs primarily intended for use with the Beomaster 8000, appeared. This new range could be easily distinguished by the new grille design, which comprised separately removable cloth panels framed in aluminium over each group of drivers. The smallest range

ABOVE: **With the grilles removed the three high-quality drive units and the two bass reflex ports can be seen.**

LEFT: **Beovox Uniphase S 55 loudspeakers were the perfect match for the Beosystem 5000. Their split grille identifies them as part of the new range of Uniphase models.**
Availability ●●●
Complexity ●●

RIGHT: **The Beovox Uniphase range included large models, such as this M 150, for use with powerful systems such as the Beolab 8000.**
Availability ●●
Complexity ●●

consisted of the Beovox S 55 and S 80, both of which were three-way pressure chamber types, the S 80 having a higher power rating and a better phase link driver. Ascending one level, the S 120 featured a forward facing bass-reflex port behind a plastic grille at the bottom of the baffle.

The S 55 was later supplemented by the cheaper S 45 (confusingly re-using the name of a previous and very successful model), a two-way loudspeaker with a permanently attached plug and cable. Minor changes were made across the range around the same time, indicated by a '.2' suffix. The final versions of the S 120 were renamed MC 120.2, though there was no real change in the specification.

Unfortunately, the new Uniphase range made widespread use of foam in the roll-edges of the drive units. This material was cheaper than the rubber that had previously been used but with time it deteriorates until in falls apart completely. At this stage either the roll-edge or the complete drive unit must be replaced to maintain the correct performance.

Beovox Uniphase S 80.2 loudspeakers were known as Beovox Perspective in some markets. These late models show the beautiful rosewood veneer that would soon disappear from the loudspeaker range.
Availability ●●●
Complexity ●●

This rather tired Beovox S 45 will need serious repairs before it can give its best again. The 'foam rot' visible here is not as the result of over-driving, it happens as the synthetic material degrades with time.

Some of the larger Beovox Uniphase models featured a protection circuit and indicator.

The new materials and new styling of the Beovox Red Line range can be seen here. These RL 60s are mounted on optional metal floor stands.
Availability ●●●●
Complexity ●●

BELOW: The Beovox Red Line RL 60.2 is a good choice for the collector as there are no foam parts to deteriorate.
Availability ●●●●
Complexity ●

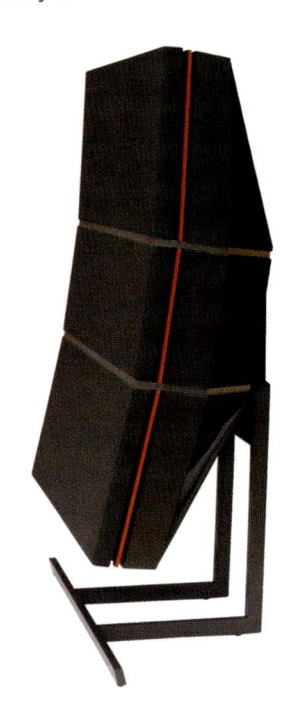

LEFT: The RL 140 (right) was quite a bit larger than the RL 60.2 (left).
Availability ●●
Complexity ●

Although generally capable of excellent performance, the Uniphase range was not particularly visually striking. Compared with the sleek new models in the rest of the range, the large wooden boxes could look out of place, limiting their appeal. A distinctive new range of loudspeakers was clearly required and it arrived in 1985 with the introduction of the Red Line series.

Named after the red line that ran around the edge of the cloth grille, the Red Line design used cast resin in place of wooden panels to make a slim and shapely cabinet. The Red Line loudspeaker could be placed either on the floor (supported by a prop at the rear), on floor stands, or on wall or ceiling brackets. The owner was encouraged to move the loudspeakers about to change the character of the sound and to make this easy, special coiled cables finished in red and black to match the cabinets, could be bought as an accessory. Initially two medium-sized models, the RL 45 and RL 60, were introduced but these were soon

1980 TO 1989: THE MICROPROCESSOR AGE   **123**

ABOVE: One potential problem area, however, is these translucent plastic bands that run around the cabinet. They are often found to be missing or broken.

RIGHT: These attractive red and black coiled RL cables are a desirable accessory for the larger Beovox Red Line models.

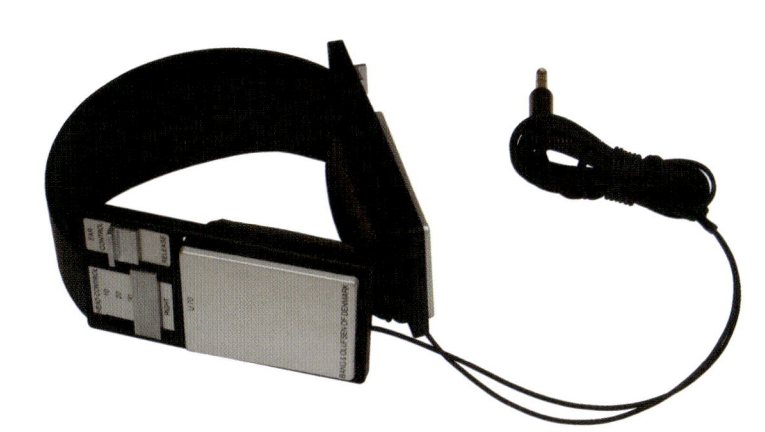

LEFT: U 70 stereo headphones were a popular accessory for the televisions and audio systems of the late 1970s and early 1980s.
Availability ●●●
Complexity ●●

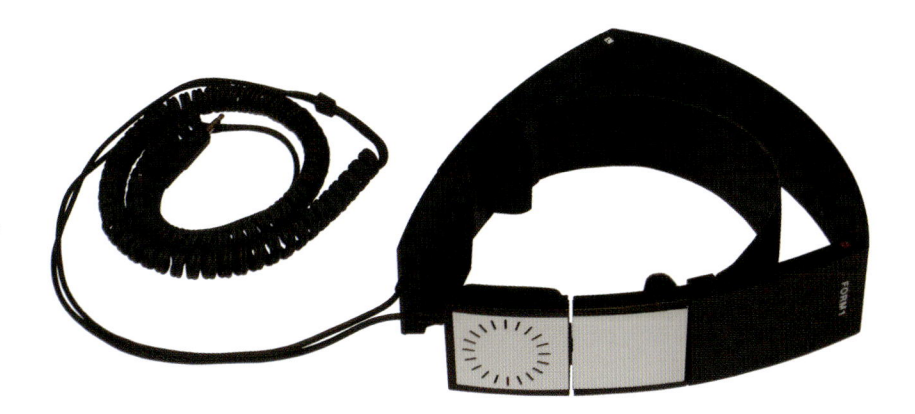

RIGHT: The Form 1 headphones – even though they look heavy and bulky, they are surprisingly comfortable to wear.
Availability ●●●●
Complexity ●

augmented by the smaller RL 35 and the very large RL 140. The RL 45 and RL 60 originally used 'auxiliary bass radiators', large floating surfaces that were driven only by the air pressure in the cabinets. These were later replaced by conventional bass-reflex ports in the RL 45.2 and RL 60.2.

Red Line loudspeaker drivers do not use foam roll edges but the bass radiators in the early models do. These deteriorate with time but repairs are not too difficult as there is no voice coil to centre. If you buy Red Line loudspeakers, try to get a pair with stands (except in the case of the RL 35, which is intended for floor-standing use only), as they are difficult to find on their own. The RL 45 is rather basic and does not give the sort of results one would expect of a loudspeaker of its size. The RL 60.2 is a much better performer and the choice model in the range.

Another long-running success that first appeared during this period was the Form 1 headphone. This replaced the earlier U 70 model and combined unusual styling with excellent sound quality. The foam pads are often found to be worn on older examples but, at the time of writing, replacement sets can still be obtained from B&O dealers.

The Move to VHS Video

Having made the mistake of backing the V2000 format, B&O bounced back and picked the JVC VHS format for its next generation of video recorders. By 1985, when B&O's first VHS machines appeared, Sony's Beta system was clearly struggling, despite some very attractive high-performance models, so the choice was an obvious one.

Of the three formats, VHS was the least technically accomplished, produced the poorest pictures and had the bulkiest cassette. However, as more manufacturers joined the VHS camp, the problems were slowly ironed out and VHS became a respectable and useful system. B&O went to Hitachi for their first VHS models: the Beocord VHS 80 and VHS 90. Hitachi were not the leading manufacturer of domestic video recorders

LEFT: **The Beocord VHS 80.** B&O responded to what the public wanted and produced a VHS video recorder, buying in the technology from Hitachi of Japan.
Availability ●●
Complexity ●●●

RIGHT: **The 'universal recording machine'.** Despite its advanced facilities, most Beocord VHS 90s were just used as high-quality video recorders.
Availability ●●
Complexity ●●●

but their VT-11 (on which the two new B&O models were loosely based) had been very well-received and had won several industry awards. The VHS 80 and VHS 90 were both front-loading machines and featured stereo sound. The VHS 90 offered a special hi-fi recording technique that used extra heads to record the sound on the tape behind the picture. This gave a level of quality comparable to compact disc and could be used either to make video recordings with excellent sound quality or sound-only recordings for which the tape speed was halved, allowing up to eight hours on a single E240 VHS tape.

The VHS 90 was marketed briefly as a 'universal recording machine' because of this feature. Priced similarly to the Beocord 9000, the VHS 90 did not cause the expected market shift and, although sound-only recording would continue to be available on Beocord VHS recorders for the next few generations, it was seldom used.

The introduction of these two stereo recorders meant that, with pre-recorded tapes as a source, there was at last a universal market for stereo television sets. The first widely distributed models were the Beovision 7802 and 8902 with 22 and 26in screens, respectively. The fitting of stereo sound to video equipment helped to form a useful link to B&O's audio products, and soon a new range of possibilities known as 'AV integration' would open up.

The first steps towards AV integration came the following year with the introduction of the Beocord VHS 91 video recorder. This simpler to operate and more compact model replaced the VHS 90 and became the top seller of the range. In simple terms, the VHS 91 was a fairly conventional hi-fi stereo video recorder, still made by Hitachi and still internally similar to some of their better models. What set it aside was that the audio connection mimicked that of a Beocord cassette deck, including the Datalink remote

LEFT: **The Beocord VHS 91 blurred the distinction between audio and video products, as it could integrate perfectly with either.**
Availability ●●●
Complexity ●●●

RIGHT: **LED indicators showed which function the VHS 91 was performing.**

The Beocord VHS 66 offered the same picture quality as the VHS 91 but had a simpler mono sound recording system.
Availability ●●●
Complexity ●●

BELOW: Fewer functions made for a simpler display.

control interface. The VHS 91 could, therefore, be operated as a second tape deck when connected to the Beosystem 5000 hi-fi system using the audio remote controls, as well as with a Beovision TV remote control in the conventional manner.

A simpler version of the VHS 91, the VHS 66, replaced the VHS 80. Re-defining the priorities of the simpler models meant that the VHS 66 was no longer a stereo machine but it did include a third video head that allowed a stable and noise-free still-picture to be produced.

Also making a brief appearance in the range was the VHS 63, a budget model based on one of the first VHS machines that Philips had designed. This model was spoiled by the use of the same Philips cabinet mouldings that were supplied to a multitude of different manufacturers to put their own brand names on. The B&O version did at least include a special remote control circuit that allowed the recorder to be operated with any Beovision TV set of the period.

Another Philips-based machine followed, the VHS 82. This featured hi-fi stereo sound and on-screen programming but its flimsy all-plastic mechanism broke easily and was not well-liked by repair shops. The styling of the VHS 82 was also uninspiring and differed only slightly from the equivalent Philips model. The cabinet top was available in a choice of four colours, black, white, silver and red, to match the Beovision MX 2000 television.

As the Beovision TV range became more sophisticated, the video recorders had to be updated to keep pace. Improved versions of the VHS 82 and VHS 91, denoted by a '.2' suffix, appeared in 1988.

The final important B&O video recorder design of the 1980s came right at the end of the decade. Known as the Beocord VX 5000, this new model was considerably more complex than anything that had come before.

Although the VX 5000 was made by Hitachi in Japan, it was designed by B&O themselves and there was no other Hitachi model that was even slightly similar. Slimmer, shallower and wider than a typical late 1980s VHS video recorder, the VX 5000 looked striking. The cabinet was so jam-packed with technology that a small internal fan was needed to keep it all cool. Basically a stereo video recorder with every conceivable convenience and quality enhancing feature, the VX 5000 had one other special facility that set it apart from all the competition. The digital video effects unit that was built into it was really a small but very powerful computer that could capture and store colour pictures, either off the air or off the tape, and then present them on the screen in a variety of different ways. The most common purpose to which this was

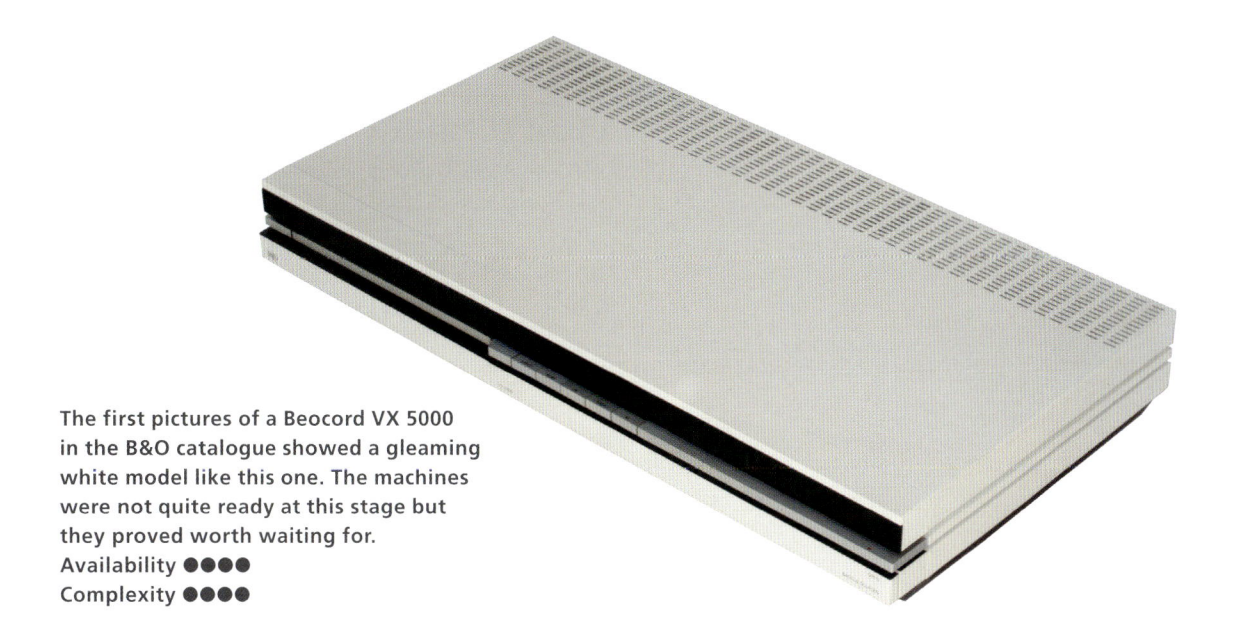

The first pictures of a Beocord VX 5000 in the B&O catalogue showed a gleaming white model like this one. The machines were not quite ready at this stage but they proved worth waiting for.
Availability ●●●●
Complexity ●●●●

put was to show a 'picture in picture', where a small image of a second TV channel (or from the tape) could be shown in the corner of the main picture. Other modes included showing up to sixteen TV channels at once by dividing up the screen into a grid, showing a frame-by-frame action sequence in a similar sixteen-frame grid, freeze-framing live TV and constructing a visual index of every recording on a tape and recording it like a contents page on the first few seconds.

The VX 5000 was very complicated and its launch was delayed several times. Those machines destined for the North American market used the Super VHS recording format, whereas those intended for Europe used normal VHS-HQ. Simpler models without the digital video effects unit were also offered in some markets.

All but the most basic functions of the VX 5000 can only be operated if it is connected to a suitable Beovision LX or MX TV set. An external VX sensor infra-red eye was offered for use with other sets but these are not easy to find.

The LX 2502/2802 or the MX 5000 are the ideal partners to use, but later models will give access to most of the functions. VX recorders are very complex and can suffer some baffling faults, many of which are caused by defective miniature electrolytic capacitors,

ABOVE: **The VX sensor.**

BELOW: **Even a small accessory like this was carefully styled and solidly built.**

of which there are hundreds. Replacements are cheap and easy to come by but the task of replacing them is trying, even for the experienced.

Stereo TV

Having got themselves comfortable with providing colour pictures, Europe's broadcasting authorities then set about adding stereo sound to television. This led to all the major setmakers, B&O included, introducing stereo TV models to receive the new services. Mention has already been made of the Beovision 7800 and 8900, which were the first two stereo Beovisions but which were not widely distributed. Stereo video tapes, laser video discs and the spread of stereo TV, meant that the next generation of models would feature in all B&O's key television markets.

LEFT: **The Beovision 7802, one of the new stereo TV models that now had mass-market appeal.**
Availability ●●●
Complexity ●●

ABOVE: **Digital tuning controls replaced the rows of tuning knobs that had been a feature of the earlier models.**

ABOVE: **The small rectangle to the left of the orange writing is a light sensor that automatically corrects the picture to suit the room lighting.**

RIGHT: **Quality cabinet finishes and attention to detail in construction put these sets firmly in the luxury class.**

1980 TO 1989: THE MICROPROCESSOR AGE

ABOVE: The new 77XX Beovisions came with a re-styled Beovision Video Terminal remote control (right), in an orange-over-grey colour scheme.

LEFT: This special all-white Video Terminal was supplied with the white versions of the Beovision 7802 and 8902 only.

RIGHT: As well as the new colour scheme, the updated terminal also featured keys for tuning and switching between the various sound modes.

The Beovision 7802 and 8902 were updated versions of the previous two sets and included new digital pre-set tuners that could store thirty-two channels, instead of the previous sixteen. This same chassis was fitted in the mono range too and these sets can be recognized by the numbers '02' at the end of the model designation. Collectively these sets are known as the 77XX series and models from this range replaced all the previous 33XX sets. A high-point in the 77XX range is the Beovision 5502, a compact and very well-equipped 20in table model, which is capable of outstanding results.

It soon became clear that the future buyers of the luxury TV sets that B&O made would demand stereo models only, so a new range was prepared. The LX series started in 1987 and comprised two sets: one with a 25in screen and one with a 28in screen. The tube used was the Philips 45AX, an FST (Flatter Squarer Tube) design that followed the design trends of the day. Despite the use of some advanced materials and technology, early 45AX tubes gave disappointing results and, therefore, the first LX sets could not give pictures that were as sharp or as bright as the previous models were capable of. Nevertheless, the LX became a popular model and developed rapidly. When NICAM digital stereo sound became available in the UK, a module was made available that could be fitted into the LX chassis, so that the new service could be received.

The original LX series models were known as the Beovision LX 2500 and LX 2800, with 25 and 28in screens, respectively. These sets were also offered without teletext or hi-fi loudspeakers as the Beovision L 2500 and L 2800. In 1988 the whole series was updated with new software that made the sets easier to operate and easier to integrate with other B&O equipment, such as the Beocord VX 5000 video recorder and the latest audio systems. The new sets could be recognized by the last digit in the model number becoming a 2, as in the Beovision LX 2802.

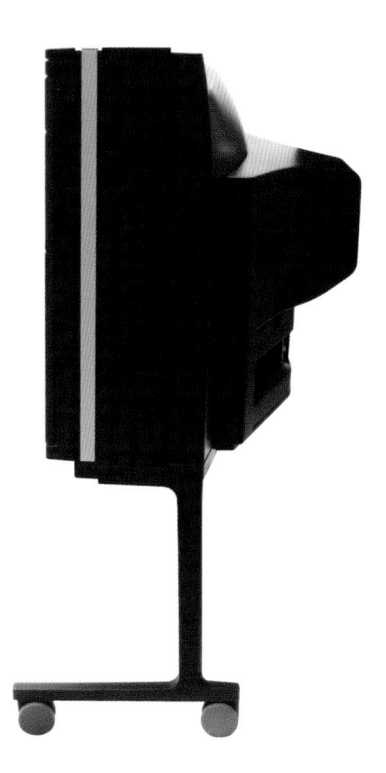

130 1980 TO 1989: THE MICROPROCESSOR AGE

ABOVE: To make the most of the new stereo broadcasts, the Beovision LX range had proper hi-fi loudspeakers. Some of the more basic models in the range lacked the tweeter.

TOP LEFT: A Beovision LX 2802 28in stereo colour television complete with a Beocord VX 5000 video recorder and trolley stand. This combination represented the top of the range offering in 1990.
Availability ●●●●
Complexity ●●

TOP RIGHT: LX quality. At a time when most television sets had bulky cabinets moulded in grey plastic, the LX range offered machined aluminium trim and fine wood inlays. This particular example is finished in ebony – a rare option.

ABOVE: As ever, the cabinet of the Beovision LX range was as slim as was technically possible.

A Fresh New Look

As well as the LX range, another important new Beovision range appeared in the mid 1980s. The MX 2000, whose cabinet was the work of David Lewis, first appeared in 1986, representing a radical departure from traditional B&O television styling. By placing the loudspeakers below the screen rather than to the sides, a more compact cabinet could be realized. Even though the 20in MX 2000 was not strictly portable, it did include hand grips along the top and underneath so that it could be moved around more easily than it was possible to do with a traditional set.

Rather than opt for the normal wooden finishes, the designers of the MX 2000 opted for high-gloss paint for the rear cover. A choice of black, white, metallic grey and bright red was offered and the Beocord VHS 82 could be ordered in matching finishes to suit. The cabinet design of the MX 2000 was greatly admired, especially the shape of the back, which was considered to be neat and visually interesting – not an easy feat to achieve with a television set. To keep the lines smooth, a tinted plastic screen was fitted over the tube face. The catalogues claimed that the tinted plastic improved the picture contrast and so the name 'contrast screen' was adopted. Despite technical objections that the tube had to work that much harder to push the picture through the dark plastic, the contrast screen was popular and remained a feature of Beovision TV sets for many years to come. A basic version of the MX 2000, the Beovision M 20, was briefly offered without the contrast screen and with an unpainted rear cover. It was not a popular choice.

Whilst the styling of the MX 2000 drew praise, the insides were not so warmly received. The slim and stylish cabinet could not accommodate the chassis of the 20in Beovision 5502 and so, lacking the resources to develop a whole new chassis for one model, B&O instead opted to buy one in. The supplier chosen was Nordmende of Germany and the chassis was known as the ICC3. Other manufacturers, notably Hitachi, had already chosen the ICC3 chassis for some of their European market models but, despite its popularity, the design was not ideal in some areas. Both performance and reliability were poor by B&O's standards and some aspects (e.g. the tuning system) were cumbersome to operate. The ICC3 chassis did not have enough power to drive the 30AX tube of the other Beovision sets, so a Toshiba tube had to be used instead; this tube did not have colours optimized for European use and so could not provide the natural hues that B&O viewers were used to. It also suffered a short life as the strain of producing a picture bright enough to pass through the contrast screen wore it out at a greatly increased rate. Later MX 2000s, recognizable by the model name being printed in white instead of orange, were fitted with a German tube made by the ITT subsidiary Videocolor, which gave better results.

Despite its technical shortcomings, the MX 2000 sold well. By 1988 Nordmende had been thoroughly absorbed into the Thomson group and the ICC3 chassis had been replaced by an updated version, the

The most striking view of the Beovision MX television is from the side, as shown here.
Availability ●●●●●
Complexity ●●

BELOW: **The styling at the rear was compromised somewhat by the bulky Philips chassis and picture tube.**

The MX 1500 re-introduced the portable to the Beovision range but the sets were not made by B&O themselves.
Availability ●●
Complexity ●

ICC5. The ICC5 was hated by the repair trade and with good reason as it was not especially reliable, was difficult to repair and originally came with a multitude of designed-in faults that required equally numerous modifications to correct. The first ICC5-based Beovision was known as the MX 3000, which was similar in appearance to the MX 2000 but was fitted with a 21in FST tube and so was slightly larger in all directions. Tube life in this model was very poor and good working examples have become unusual indeed at the time of writing.

The ICC5 chassis was capable of driving picture tubes up to 28in in size and so a larger Beovision MX model was introduced to make use of this. The styling scaled-up well, resulting in the MX 4500, which was also a strong seller despite its poor picture and dubious reliability. A later model, the MX 5000, included updated software to give optimum compatibility with the VX 5000 video recorder. This model introduced a popular innovation, the motorized rotating stand or table base.

To complement the bigger MX series, a small 'portable' model, known as the MX 1500, was also introduced. Making its debut late in 1987, this model had a 15in FST tube and mono sound. It certainly looked smart and could be operated with the same remote control unit that the larger Beovision sets used. However, inside was a Philips chassis known as the CP90, which although better than the ICC5 was still not up to the levels of performance and reliability that the B&O buyer would expect. The complete MX 1500 was assembled in a Philips factory in Italy and only the Beolink 1000 remote control it came with was made by B&O.

A Gentle Entry into Compact Disc

The compact disc digital audio format made its world debut in 1981 and became available in Europe during the following year. Developed jointly by Philips and Sony, it combined the convenience of the cassette and the sound quality of the LP record. Although compact disc was not the first digital audio format, it was the first that was suitable for home use and so aroused a lot of interest at the time. The silver optical discs came from the fallout of the failed Philips Laser Vision video disc format. Laser Vision was similar to DVD in many ways, though at 30cm in diameter the discs were considerably larger. However, the public were not receptive at the time to a non-recordable video disc format, preferring instead video cassette recorders, which proved far more versatile for a similar outlay. Sony brought their digital recording expertise to the project, adapting their PCM digital audio format and its attendant error correction software from its previous role, as part of a professional recording system used with modified industrial video recorders, to a domestic playback-only role based on the Philips optical disc.

Early compact disc players were very complicated and required considerable precision manufacturing expertise to produce. While most of the familiar Japanese brands managed to produce a player of some kind within a few years, the Europeans struggled and, in most cases, simply opted to buy the technology and the major components directly from Philips or one of the Japanese firms. This is the route that B&O took with their early models.

Most manufacturers sold compact disc mainly on its sound quality. This was nothing new for B&O, as for some years the top line Beograms and Beocords had been capable of similar levels of performance, but compact disc also offered simplicity of operation and convenience. The freedom from mechanical controls and the ability to find any track instantly and unfailingly suited the B&O long-term objective of offering perfect sound at the press of a single key perfectly. Compact disc was also a natural part of a remote control multi-room system.

The first B&O compact disc player appeared in the 1984 range catalogue. Named the Beogram CD 50, it was designed to match the Beosystem 5000 in every respect. An apparently complete and working machine was shown in numerous photographs but the text made clear that the CD 50 was not yet available to buy. Showing it did put the minds of prospective Beosystem 5000 owners at rest, knowing that they could buy the rest of the system without fear of it becoming obsolete.

The CD104 was a typical early CD player from Philips. It was used by many smaller manufacturers, B&O included, as a basis for their early compact disc models.

The Beogram CD 50 that appeared towards the end of the following year differed in numerous minor details to the one shown in the first catalogue, but the concept remained unchanged. It was clearly part of the Beosystem 5000 and it integrated perfectly in both technical and aesthetic respects. Even the name was carefully chosen: 'CD 50' had the same number of characters as '5000', so the model names on all the 5000 series components were all of a matching length.

Based on an Aiwa design, the CD 50 used a three-beam laser pickup mounted above the disc tray. For this reason, the compact discs had to be loaded label-side down, something that would cause a lot of confusion in the following years. Another technical quirk could be found in the digital to analogue converter (DAC) stage; a single high-precision sixteen-bit DAC chip was used that was switched rapidly between the left and right channels to give a stereo output. Complex analogue reconstruction filters followed the switching stage, which worked so well that even the most discerning listener could not tell what was going on inside.

As delivered, the CD 50 had only three controls: one to open the drawer, one to change the display and one to perform the standby/play/skip track function. These were enough to make the machine usable but for full access to the programming functions, a Beomaster 5000 and its Master Control Panel would usually be needed. As an alternative for owners of

LEFT: **The Beogram CD 50 had an unusually styled disc drawer into which the discs were loaded label-side down.**
Availability ●●●
Complexity ●●●

ABOVE: **With the drawer closed, the Beogram CD 50 looked just like the other Beosystem 5000 components.**

LEFT: **The Beogram CD 50 was part Danish and part Japanese. The cabinet and brown circuit board on the right are the main B&O-designed parts.**

other systems, a remote control receiver could be built into the compact disc player by the dealer and then the machine could be used with its own special remote control called the CD Terminal.

Whilst the Beogram CD 50 was the first B&O compact disc player to be shown, it was not the first to become widely available. Earlier in 1985, before the CD 50 appeared, another radically different model was launched. Known as the Beogram CD X, this player was made for B&O by Philips. It was built using an odd collection of Philips components: the main chassis came from the Philips CD100 (the first Philips compact disc player), whilst the optical and electronic parts came from the CD104, a popular model that was very well-respected at the time. The casework was rather larger than either of the Philips machines on which the CD X was based, but was strikingly styled. On pressing the 'Open' button, the clear plastic and aluminium lid glided upwards and the disc tray (which looked like the platter of a miniature record

The CD Terminal is a rare accessory for the Beogram CD 50. A receiver module has to be built into the CD player in order for it to work.

RIGHT: **Open and ready to receive a disc. The silver 'turntable' was in fact fixed and did not rotate.**

LEFT: **The beautiful Beogram CD X broke with the trend of black boxes with drawer loading. The controls were touch sensors on a glass panel.**
Availability ●●●●
Complexity ●●

player but in fact did not rotate) lifted, ready to receive the disc. The controls were touch-sensitive and elaborate programming facilities were included. The CD X was designed to be used with any Bang & Olufsen audio equipment (hence the 'X' in the model name) but its styling and colouring suggested that the Beosystem 2000 and 3000 were the ideal partners. The red line running through the disc loading area, in particular, was a common styling theme with these models. Along this line in the CD X was written 'Laser Optical Turntable', a description that, although accurate, did not stick in the public's mind and was not heard of again.

In common with some other early Philips compact disc players, the Beogram CD X used a dual fourteen-bit decoder. One may think that not decoding all the information from the disc would result in inferior sound quality, but this was far from the case. Philips clearly knew exactly what they were doing as the CD X was capable of magnificent performance, sounding smooth and civilized when compared to the rough and edgy presentation of some early Japanese models.

With the Beogram CD 50 and the Beogram CD X, B&O had two excellent compact disc models in their range that had required relatively little development work on their part. Both sold well, were solidly made and remained firm favourites with B&O owners throughout the CD era. However, in such a fast-moving field it was not surprising that new models would quickly arrive and replace them. The CD X was replaced by two models: the Beogram CD 3300 and the Beogram CDX 2. These were two versions of basically the same machine. The CD 3300 was styled and

LEFT: **The Beogram CD 3300. The styling is similar to the previous Beogram CD X but the internals are very different.**
Availability ●●●
Complexity ●●

BELOW LEFT: **The elegant CD tray of the Beogram CD X was retained, but now it was surrounded by silver-painted plastic and the 'laser optical turntable' script was gone.**

BELOW RIGHT: **The large graphics and grey finish are very much of the period. Note that there is no mains switch as this model is designed to be remote controlled.**

RIGHT: **The Beogram CD 5500 followed the existing styling pattern but the new drawer design meant that the front panel no longer needed to open.**

BELOW: **A view inside the Beogram CD 5500. Not only is it much neater than the previous CD 50 model, it is also completely designed by B&O using Philips components.**

ABOVE: **One of the most elegant features of the Beogram CD 5500 was the disc-loading drawer. It was machined from solid aluminium and moved briskly and silently. This is an early model which lacks a second groove in the tray for locating 8cm CDs.**

coloured to match the Beosystem 3300 (the successor to the 2000 and 3000 ranges) and included Datalink, whereas the CDX 2 continued the original CD X theme and retained the advanced programming options. Both were based on the sixteen-bit Philips CD150 series models but retained the CD100 chassis casting. The same basic mechanism and electronics were also used in Bang & Olufsen's first compact disc music centre, the Beocenter 9000.

The Beogram CD 50 was replaced by the Beogram CD 5500, which can be regarded as the first true B&O compact disc player. This also used the Philips CDM optical deck and sixteen-bit chipset, but the printed circuit panels and supporting electronics were of B&O's own design. Thoroughly rationalized power supply and analogue output circuitry were the key features of the Beogram CD 5500 design; these accounted for the outstanding performance that this model was capable of. The analogue filters achieved the required performance without the use of inductors and this, along with the use of carefully selected samples of the famous Philips TDA1541 DAC chip, made for very low distortion and an exceptionally smooth, refined sound.

The final new compact disc player of the 1980s arrived in 1989. The Beogram CD 4500 was designed as part of the Beosystem 4500 (the final version of the line that had started with the Beomaster 1900 of 1976) and this used conventional Philips circuitry laid out on B&O circuit boards. Laying out the boards themselves gave the designers the flexibility to house the machine in the same cabinetwork as the Beocord 4500 cassette deck and, therefore, make it a better styling match with the rest of the system than the Beogram CD 3300, which it replaced. It also allowed the deck suspension to be made adaptable for vertical wall mounting along with the Beomaster and Beocord, another first.

The early compact disc models have proved themselves durable and are an excellent addition to a system of the correct period. The CD X, in particular, has established a good record as a dependable machine, although faults do occur, so test carefully. The story that CD pickups do not wear out like LP styli was one of the many myths that came about during the early years of compact disc. In truth, with use the laser diode gradually loses power until the disc can no longer be read, meaning that a replacement is needed. Faulty CD 50s should be approached with care for, unlike the Philips-based models, no technically similar machines were widely distributed under more mainstream brand names.

Rationalizing Remote Control

Although B&O had not been the first to introduce remote-controlled TV and audio products, they quickly realized that remote control could add a new aspect to the enjoyment and convenience of operating their equipment. The remote control system for television sets had developed logically to the extent that, by the beginning of the 1980s, all the models used the same remote control unit, which could also operate all the suitably equipped video cassette recorders. The situation was not so well organized for audio, as each range had its own type of remote control, all incompatible with each other and, in some cases, unable to control the complete system. There was almost no interaction possible via remote control between audio and video products, which meant that owners who had decided to opt for B&O for all their equipment, ended up with at least two incompatible and differently styled remote control units – not a desirable state of affairs.

The answer came in 1987. A new system of infrared remote control appeared in the range in three important new models based around a range of slim and attractive remote control handsets. Initially these were known as the A Terminal, the V Terminal and the AV Terminal, and were for audio, video and all models, respectively. This new range was different to what had come before in one important respect: because the rate at which the new remote controls sent information was much faster than was common for a typical remote control system, far more information could be transferred at each key press. This in turn made it possible, not only to send a command (for example 'play' or 'programme 1') but also to specify which piece of equipment it was for (audio tape, television, turntable and so on). This offered the possibility of producing a simple remote control terminal with a manageable number of keys that could operate any audio or video source.

Initially, the V Terminal was supplied with the Beovision LX TV range and the A Terminal was included with the Beosystem 5500 and the Beocenter 9000. The Beosystem 5500 was a redesigned Beosystem 5000. The Beomaster and Beocord were substantially altered, the Beomaster to accommodate the new remote control system and the Beocord to include auto-reverse, automatic Dolby recognition and automatic recording level control. The turntable was altered in name only (from Beogram 5005 to Beogram 5500) and to start with, the Beogram CD 50 compact disc player was carried over unchanged, to be replaced by the Beogram CD 5500 when it became available during the following year. The other new audio unit, the Beocenter 9000, was an important new compact disc music centre.

The AV Terminal was a desirable accessory for those who bought a complete audio-visual system and included a small, simple LCD display to show which

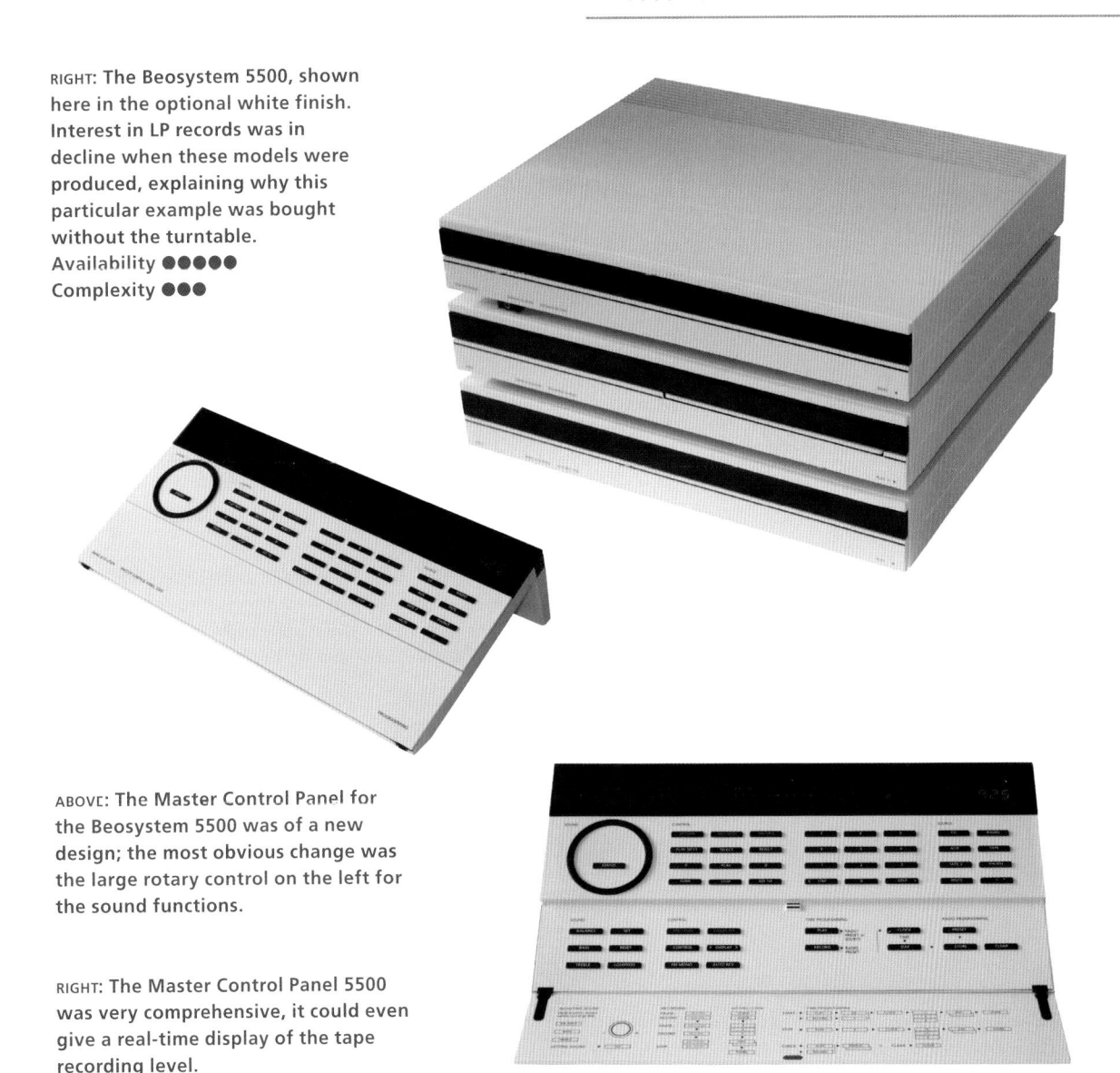

RIGHT: **The Beosystem 5500, shown here in the optional white finish. Interest in LP records was in decline when these models were produced, explaining why this particular example was bought without the turntable.**
Availability ●●●●●
Complexity ●●●

ABOVE: **The Master Control Panel for the Beosystem 5500 was of a new design; the most obvious change was the large rotary control on the left for the sound functions.**

RIGHT: **The Master Control Panel 5500 was very comprehensive, it could even give a real-time display of the tape recording level.**

mode (audio or video) was selected. As well as being useful for controlling equipment directly, all the new remote control units could be used with the latest version of the multi-room system, which had been redesigned to make best use of the possibilities that the new remote controls offered.

The three new remote controls were a move in the right direction but the ideal of having a single unit that was suitable for all products was not achieved in full until 1988. The Beolink 1000, a simplified replacement for the AV Terminal, was without doubt the best and most rational remote control unit offered by any manufacturer at the time, and remains to the present day an object lesson in the ergonomics of keypad design. It looked simple and was easy to use, yet it could operate all the main functions of all the major audio and video products in the range.

By 1989 all the other remote control units in the range had been deleted, with the exception of the Master Control Panel 5500 and a simple unit for the Beosystem 3300 (which could be replaced by the Beolink 1000 anyway).

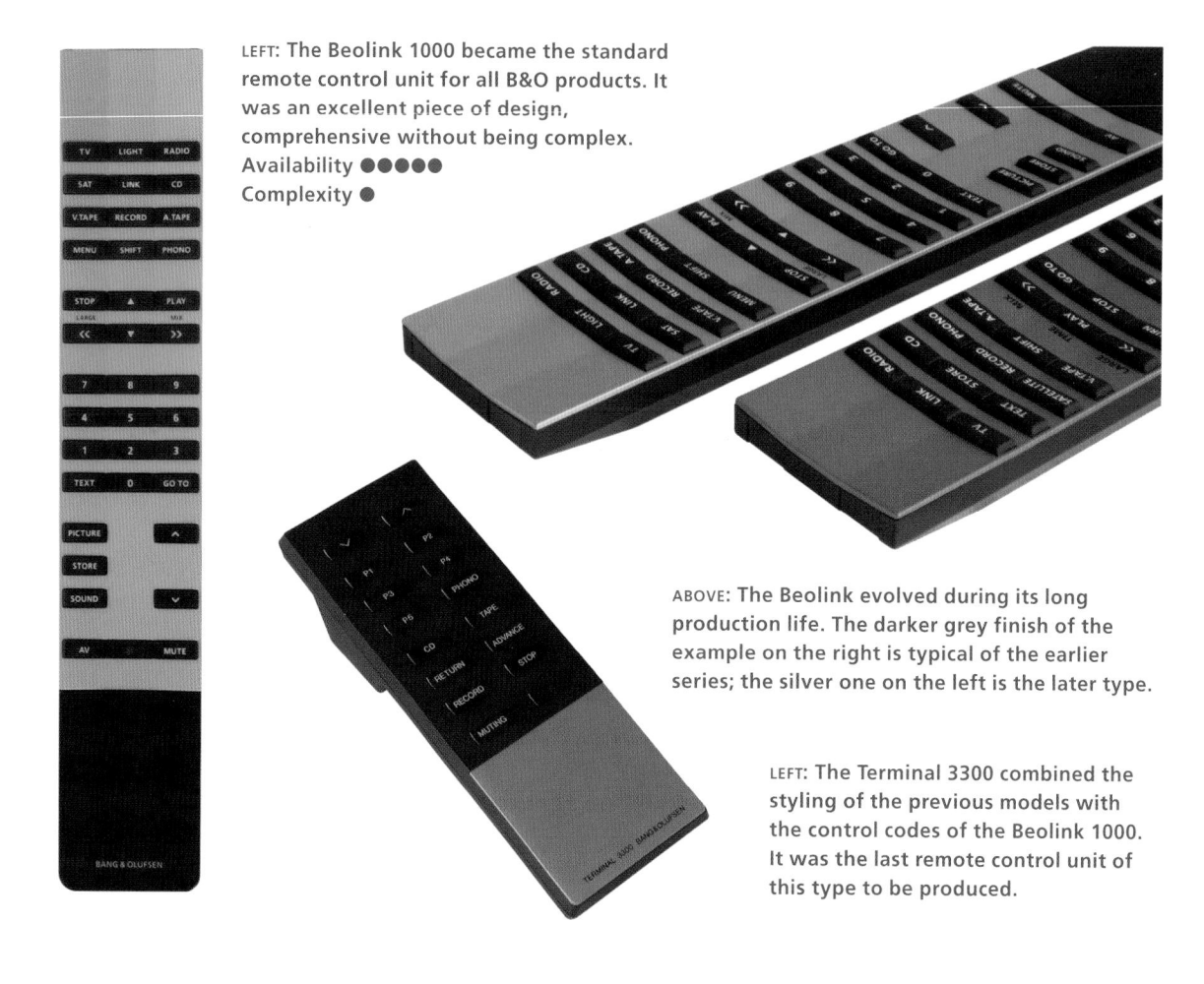

LEFT: **The Beolink 1000 became the standard remote control unit for all B&O products. It was an excellent piece of design, comprehensive without being complex.**
Availability ●●●●●
Complexity ●

ABOVE: **The Beolink evolved during its long production life. The darker grey finish of the example on the right is typical of the earlier series; the silver one on the left is the later type.**

LEFT: **The Terminal 3300 combined the styling of the previous models with the control codes of the Beolink 1000. It was the last remote control unit of this type to be produced.**

The Beolink 1000 instantly became a design classic and the system of control codes it introduced remained the only standard that B&O used right up to the time of writing.

A Big New Sound

The deletion of the range of the final version of the Beomaster 6000 and its attendant sources left a potentially embarrassing problem for Bang & Olufsen. Without the 6000, the new Beosystem 5500 naturally rose to the top of the range and, despite its general all-round competence, it was a full 20W less powerful than the previous top-line series. The answer came not from a new receiver but from a new type of loudspeaker: the Beolab Penta.

The Beolab Penta was a large column loudspeaker with four bass units, four midrange units and a centrally mounted tweeter. Its tall, elegant cabinet was finished in gleaming stainless steel and its small footprint meant that it was easy to accommodate for such a large model. However, what really made the Beolab Penta different from previous B&O loudspeakers was the amplifier at its base. Each loudspeaker included an amplifier, which, although compact, could deliver 150W from either the loudspeaker output of a smaller amplifier or from a line level signal. Using Beolab Pentas with the Beomaster 5500 formed a system with the same power output as the Beomaster 8000 was capable of, with the added advantage that the ampli-

1980 TO 1989: THE MICROPROCESSOR AGE   **141**

ABOVE: The controls for the powerful amplifier at the base of the Beolab Penta are hidden behind this folding panel.

LEFT: The Beolab Penta. Moving the power amplifiers from the main unit into the loudspeakers had many technical advantages and gave the designers greater freedom in the way that the source units could be styled.
Availability ●●●●
Complexity ●●

fier and loudspeaker were perfectly matched and connected together only by a very short length of cable.

When used with either the Beomaster 5500 or the Beocenter 9000, the Beolab Pentas revealed another surprise: an illuminated display panel located between the two cloth frets, which showed which source and track was playing and at what volume level – a useful addition, especially if the loudspeakers were located in a room distant from the main system. The information for the displays was sent to the loudspeakers along with the sound through special cables with four-pin plugs. The system was known as Speaker Link.

The Penta loudspeakers could be bought without the amplifiers or displays for use with powerful receivers like the Beomaster 6000 or 8000, in which form they were known as the Beovox Penta. The amplifier was also available separately to give a boost to small systems used with large loudspeakers and was given the model name Beolab 150.

The digital display fitted to the Beolab Penta could show the volume setting and which source was playing. This type of LED display was used in the original Mk.I model.

142 1980 TO 1989: THE MICROPROCESSOR AGE

This style of display was used in the Mk.II Beolab Penta. It uses LCD panels and shows more information than that of the previous model.

The Mk.III Beolab Penta used a green dot-matrix type display.

ABOVE: The base of the Beovox Penta.

LEFT: The Beovox Penta, shown on the left, was shorter than the Beolab model because the amplifier had been removed. The cabinets were otherwise identical.
Availability ●●
Complexity ●

These loudspeakers should be inspected carefully before buying. The amplifiers receive little ventilation and, in some installations, are never completely switched off, so faults are not unknown. Crackling or hissing noises in the loudspeakers, even with no signal input, are common signs that all is not well. Another sign of impending problems is that the indicator light at the bottom of the column turns orange (rather than the usual red or green). Minor knocks and dents in the cabinet sides are par for the course and there are very few perfect examples about, but that does not mean that one has to accept a pair with deep scratches, peeling edges or missing parts. The loudspeaker units themselves have proved durable but the roll edges of the mid-range units can rot away. There are eight of them and they are surprisingly expensive to replace.

LPs Out, Compact Discs In

Throughout the first half of the 1980s, the compact disc player had, in the main, stayed at the top end of hi-fi, a sophisticated and costly device that played expensive discs in a way that seemed almost magic. However, as manufacturing techniques improved it became possible to reduce the cost and complexity of the compact disc player to the point where it could be included in a music centre. The first compact disc music centres from other manufacturers were simply re-worked 'midi systems' (a small one-piece unit with the appearance of a stack system) with a compact disc player added to the bottom, the cost of which usually exceeded that of the whole of the rest of the system. Although the record player was in decline, few manufacturers were willing to remove it completely.

The size of the LP record had limited designers to the styles that practical music centres could be made in. Once the compact disc started to push records aside, it was realized that there was no longer a need to accommodate a 30cm circle into the cabinet somewhere and a new look was established. Low, long, sleek and wide, the new compact disc music centres looked striking and modern. In the UK, Hitachi offered their Opus 1 model to great acclaim, but even this looked crude compared to B&O's new design, the Beocenter 9000.

Introduced for the 1987 season, the Beocenter 9000 was a striking sculpture in black glass and brushed aluminium. Its simple lines, vaguely reminiscent of the Beocenter 4000 of 1978, were completely uncluttered, as there were no knobs, sliders, keys or buttons. Every function was operated by touch-sensitive areas on the glass panels. To load a cassette or a compact disc, all one had to do was to touch the glass and the covers at the sides glided open. Once closed, no moving parts were visible; the only way one could tell that a source was playing, was by its perfect sound. The Beocenter 9000 was a very advanced unit, featuring fully programmable operations of all the sources and a comprehensive electronic timer. Both

The beautiful Beocenter 9000 was the first of a new generation of combination units that offered a CD player instead of a turntable.
Availability ●●●●●
Complexity ●●

the cassette recorder and the compact disc player could be programmed to play the tracks in any order and the radio could be pre-set with up to twenty stations from any waveband. Such versatility could have made the Beocenter 9000 almost unusably complex but for some very clever design – the electronic covers. The electronic covers were a system that meant that only the controls relevant to the function at hand were visible. In practice, this was done by making the touch-sensitive controls visible only when illuminated by red lights from beneath. So, instead of a console of many confusing controls, only a few possible options are visible at any one time. For example, if the radio was selected, then only the simple functions of station selection and volume control could be seen (or used), the tape recorder and compact disc player were in effect invisible, with the exception of the 'record' function if a suitable cassette was loaded. Switching to compact disc immediately re-configured the panel, as would selecting any of the other sources. A further depth of control for each function was hidden behind a common control simply labelled 'programming', which could be used or ignored as the owner chose. Full remote control was part of the standard package using the A Terminal (and later the Beolink 1000) and the Beocenter 9000 was also prepared for multi-room operation.

Despite its apparent complexity, the Beocenter 9000 was not that difficult to produce. Only the computer control system was really new, as the radio and pre-amplifier came from the Beomaster 5500, the tape deck circuits and mechanism from the Beocord 5500, the complete compact disc player from the Beogram 3300 (itself based heavily on a Philips design) and the power amplifier from the Beomaster 3300. B&O could, therefore, genuinely claim to offer hi-fi separates performance with music centre convenience; a truly compelling package.

The cassette section, in particular, offered some very novel features to help the owner make good-quality recordings with minimal knowledge or skill. Firstly, it included auto-reverse and automatic level control. The two sides of the cassette were treated as one single long length and so, without difficulty, the user could assemble long recordings without having to worry about running out of tape or matching the sound levels. Both these automatic functions could be over-ridden if necessary and the machine used in 'fully manual' mode. Most technically interesting was the automatic Dolby NR selector. Non-technical users found the Dolby noise reduction systems of some recorders complicated and sometimes suffered poor-quality playback if the same settings were not used during playback that were in place when a recording was made. The Beocenter 9000 solved this problem by recording extra 'hidden' signals onto the tape that told the microcomputer in the machine which Dolby setting (B type or C type) was used to make the recording and set the decoder to the correct mode during playback automatically. This clever system was offered to other manufacturers under license but was not widely taken up, nor did it feature on pre-recorded tapes, which was a great shame.

The Beocord 5500 cassette deck also included all the technological refinements of the Beocenter 9000 cassette section along with extra manual control options.

The designers had been brave to exclude a record player from the main unit but it was realized that any serious listener was likely to still have a large record collection, so the Beocenter 9000 included a connection for a turntable. The model recommended initially was the Beogram 3000, later on a special type called the Beogram 9000 would be offered instead. The Beogram 3000 is quite common, as it was also used with other equipment but the Beogram 9000 is quite unusual.

The Beocenter 9000 is something that everyone seems to like, so good ones are seldom that cheap. Advancing years have, in some cases, caused the computer control systems to act strangely and some of the compact disc players are coming to the end of their lives too. This is a very complex unit, it is effectively a whole separate system fitted into one quite small box and hinging open the cover reveals a great deal of circuitry and mechanism. It comes as no surprise then that when attention is needed, it can be involved or expensive, so it pays to start with as good an example as can be found. Like the Beosystem 5000, they are plentiful, so be selective.

CHAPTER FIVE 145

1990 TO 2000: THE SURVIVAL YEARS

The early years of the 1990s were difficult times for European industry. If the 1980s had been a party, then the 1990s represented the hangovers and clearing up that followed, bringing with them economic difficulties and tough trading conditions in key markets. B&O had started to struggle in the closing years of the 1980s and they were not alone. Many firms in a similar position simply disappeared never to be seen again, whilst others became just names that were cynically applied to low-brow equipment.

B&O's response was pragmatic – they simplified their range and cost-reduced those items that remained. The previously mentioned Beosystem 5500 was a good example of this process, being simpler to produce than the original Beosystem 5000 but still saleable at a premium price. In some ways, however, the cost cutting went a little too far with some models ending up cheaply finished with medium-grade black plastic mouldings and thin metal trims, where in the past far more solid and sturdy parts would have been used. Some expensive but little-used technologies were also abandoned; for example, the digital picture processor on the Beocord VX 5000 video cassette recorder and the real-time tape counter of the top-line audio cassette decks. The 1990s also saw B&O's withdrawal from the serious turntable market, when the Beogram range was reduced to just one basic model, which was offered in a range of cabinet styles to suit the various audio systems in the range. In all forms it was rather overpriced for the performance it was capable of and this, along with a lull in interest in the LP format in general, led to poor sales. This can be a problem for today's collectors, if they want to assemble a complete system.

An interesting new product line emerged in 1990: the B&O telephone. The first model was known as the Beocom 2000, which took the form of an attractive desk-top instrument with a coloured keypad and a simple display. Various colours were available in either bold or pastel shades and, even though there were no

The Beocom 2000 brought hi-fi quality to the domestic telephone with its carefully engineered microphone and loudspeaker.
Availability ●●●
Complexity ●

146 1990 TO 2000: THE SURVIVAL YEARS

The Beocom 2000 was available in many different colours. The keypad of later models took on more subdued pastel tones.

clear styling cues from any recent B&O equipment, the Beocom 2000 fitted perfectly into the range. The telephone range would expand through the 1990s and rapidly became a popular sideline.

Also introduced in 1990 was the Beovox Cona, a passive subwoofer designed to work with the Beovox CX range of miniature loudspeakers. The Beovox Cona was designed for music lovers rather than film fans and so did not shake the floor; it just gave the sound a richer, fuller nature in a very subtle manner.

Revised Audio Systems

New technology and a desire to 'freshen up' the audio range led to revisions across the range for the 1990 season. The most important change that was made was the addition of a new type of loudspeaker connector, Power Link.

The Beovox Cona. Hidden discreetly away this model can give a real boost to the Beovox CX range.
Availability ●●
Complexity ●●

Power Link was a way of connecting amplified loudspeakers that used a single small plug to carry all the signals and control data along one cable. As the cable carried both the left- and the right-hand signals, the loudspeakers could be 'daisy-chained' together. A small switch on each loudspeaker selected it as either on the left- or the right-hand channel. In its full implementation, Power Link also carried data for display panels mounted on the loudspeakers, as well as control signals for on/off switching and muting. The first Power Link loudspeaker was the Beolab Penta 2, which was otherwise very similar to the original model. The Penta amplifier could still be bought separately and in Power Link form was now known as the Beolab 200; but, despite what the name suggested, the power output remained at a maximum of 150W. Also available were two new wall-mounted panel-type models: the Beolab 3000 and the Beolab 5000. These were strikingly styled with bold cloth materials and polished metal panels, although behind all this they were of conventional design. The amplifier/display units were fitted beneath the loudspeaker and were of a far simpler design than those used for the Penta range, the key component being a hybrid chip amplifier made by Sanyo. In order to improve the performance, the circuit included stages whose bass response changed according to the level of the incoming signal. The idea behind this was to boost the bass at low volumes to compensate for the change in the ear's sensitivity that occurs under these conditions. This function replicated that of the loudness control found on some amplifiers, with the disadvantage that it could not be turned off to give the loudspeakers a linear response. Nevertheless, such systems would become more and more commonplace in the Beolab loudspeakers that would follow, even though they offended the purist.

The panel loudspeakers were also available without the amplifiers as the Beovox 3000 and 5000, presumably leaving the listener to make the necessary bass corrections manually. The Beolab 3000 proved popular and would enjoy a long production life. Not so the Beolab 5000 and Beovox 5000, which had disappeared from the catalogues in little over a year.

The Beolab 3000 reintroduced the panel loudspeaker to the range. The amplifier is housed behind the black and silver panels at the bottom.
Availability ●●●
Complexity ●●

The Power Link audio systems were all revised versions of the previous season's offerings. The Beosystem 6500 replaced the 5500, the Beocenter 9500 replaced the 9000 and the Beosystem 4500 continued the 1900 line, replacing the 3300, which had been available only for a year.

As well as Power Link, these models also all included a new type of two-way infra-red remote control. Remote control systems of this sort had been seen before, but at last the control codes had been made completely compatible with the Beolink 1000 format and so became universal across the whole range, including the video products. In 1990 this was not of any great value but in 1991 a new range of remote control terminals was introduced that made full use of the systems.

148   1990 TO 2000: THE SURVIVAL YEARS

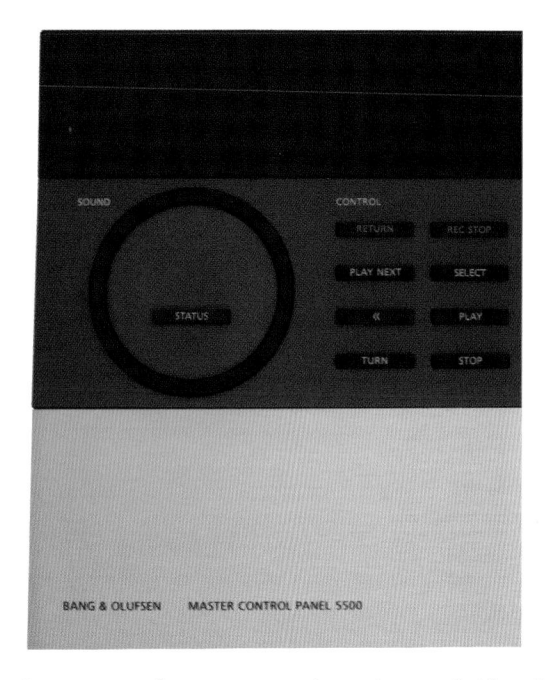

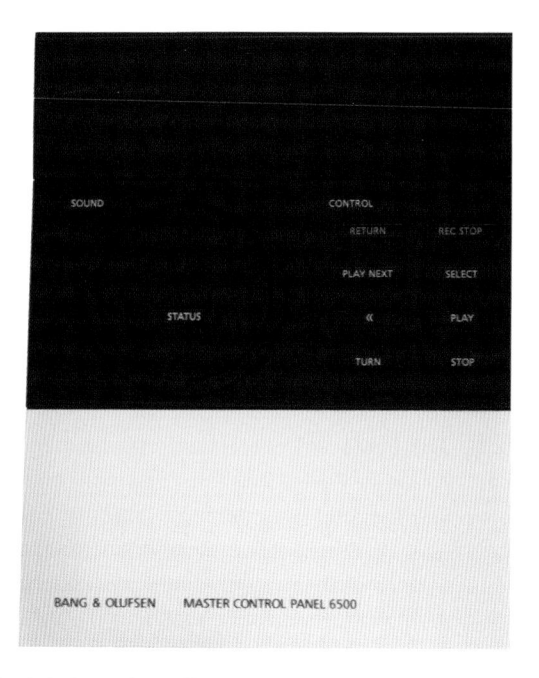

These views of Master Control Panels 5500 (left) and 6500 (right) show the differences between colour and finish of their respective Beosystems.

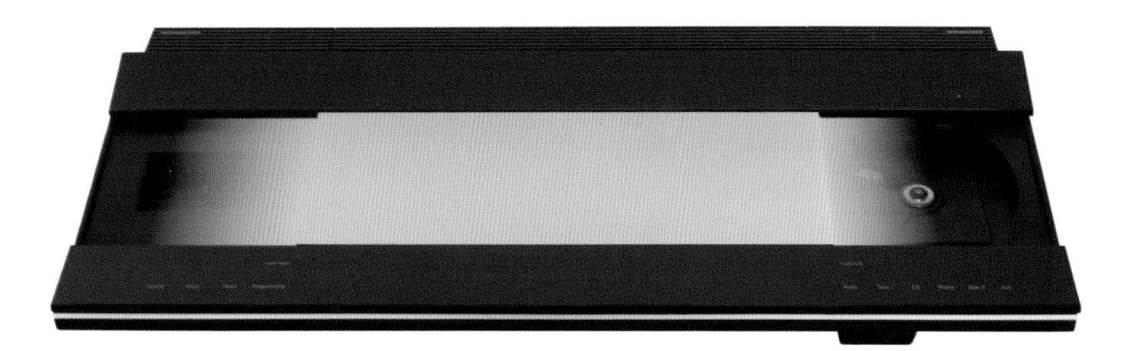

The doors of the Beocenter 9500 glide open in response to the lightest touch. This model was one of the best B&O products of the 1990s.
Availability ●●●●●
Complexity ●●

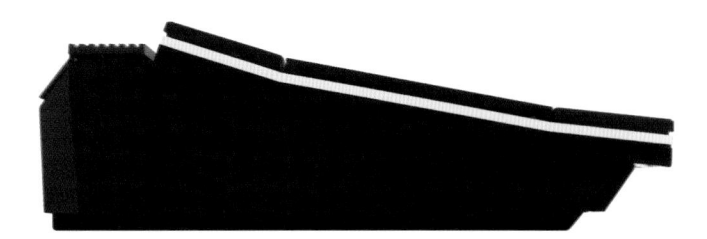

LEFT: Typical B&O attention to detail made the Beocenter 9500 attractive in profile too. The thin white line replaced a wide bronze-coloured strip that was fitted to earlier models.

1990 TO 2000: THE SURVIVAL YEARS **149**

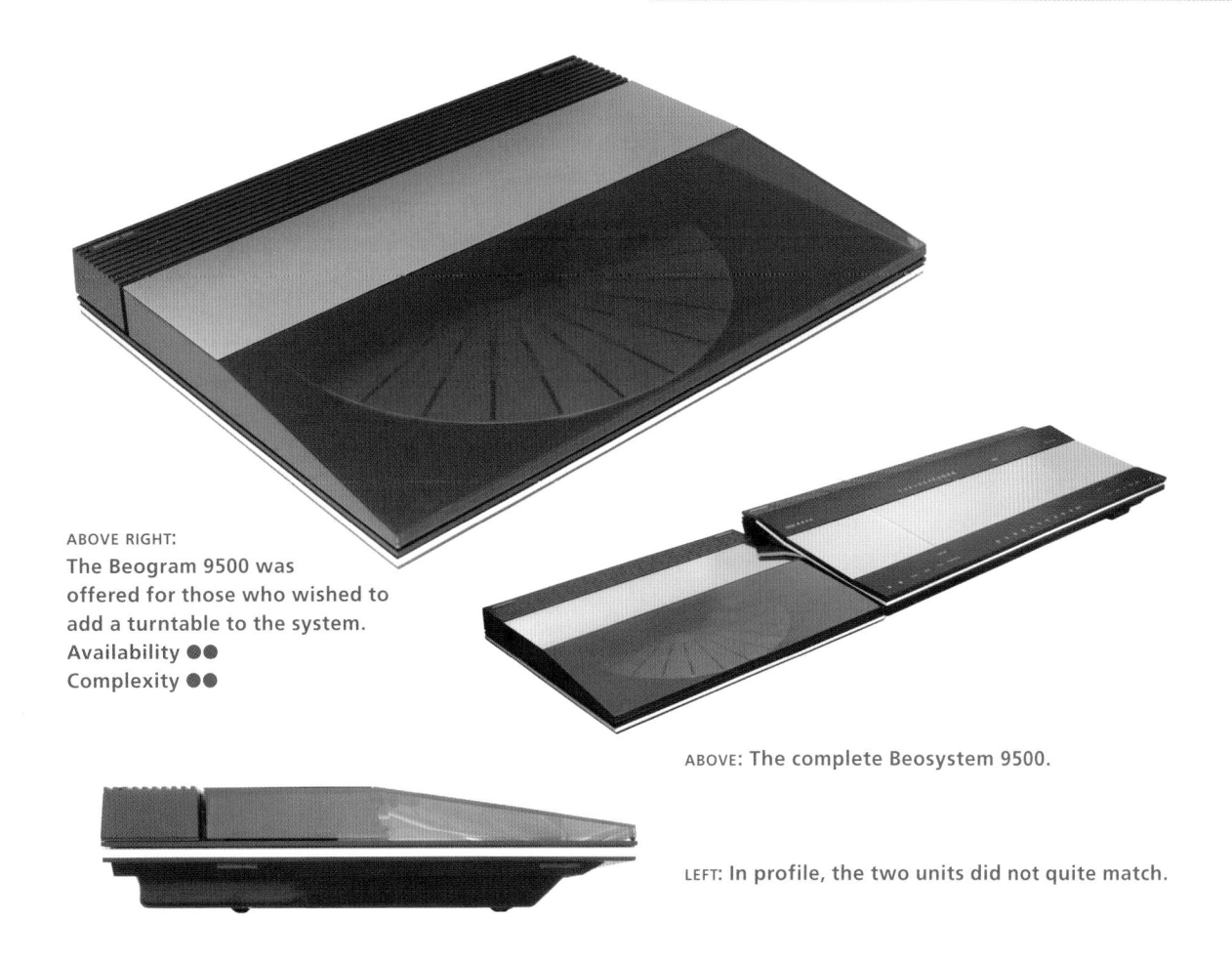

ABOVE RIGHT:
The Beogram 9500 was offered for those who wished to add a turntable to the system.
Availability ●●
Complexity ●●

ABOVE: The complete Beosystem 9500.

LEFT: In profile, the two units did not quite match.

The Beolink 5000 was about the same size as the Beolink 1000 but included a double-sided transparent LCD window that guided the user through the programming of the system that it was being used with and also showed the levels of volume, treble, bass and so on, that had been set.

The Beolink 7000 was a larger and more complex unit that used a touch-sensitive LCD screen in conjunction with a touch-sensitive illuminated panel to offer complete control over a whole system. It replaced the Master Control Panel 6500 as the remote control for the Beosystem 6500, but it was equally applicable to the latest television sets and video recorders. The Beolink 7000 was considerably smaller than the Master Control Panels but required more power. With no space to house the four large torch batteries that had been used previously, another

The Beolink 5000 was certainly impressive but it is also quite difficult to use and rather fragile.
Availability ●●●
Complexity ●●●●

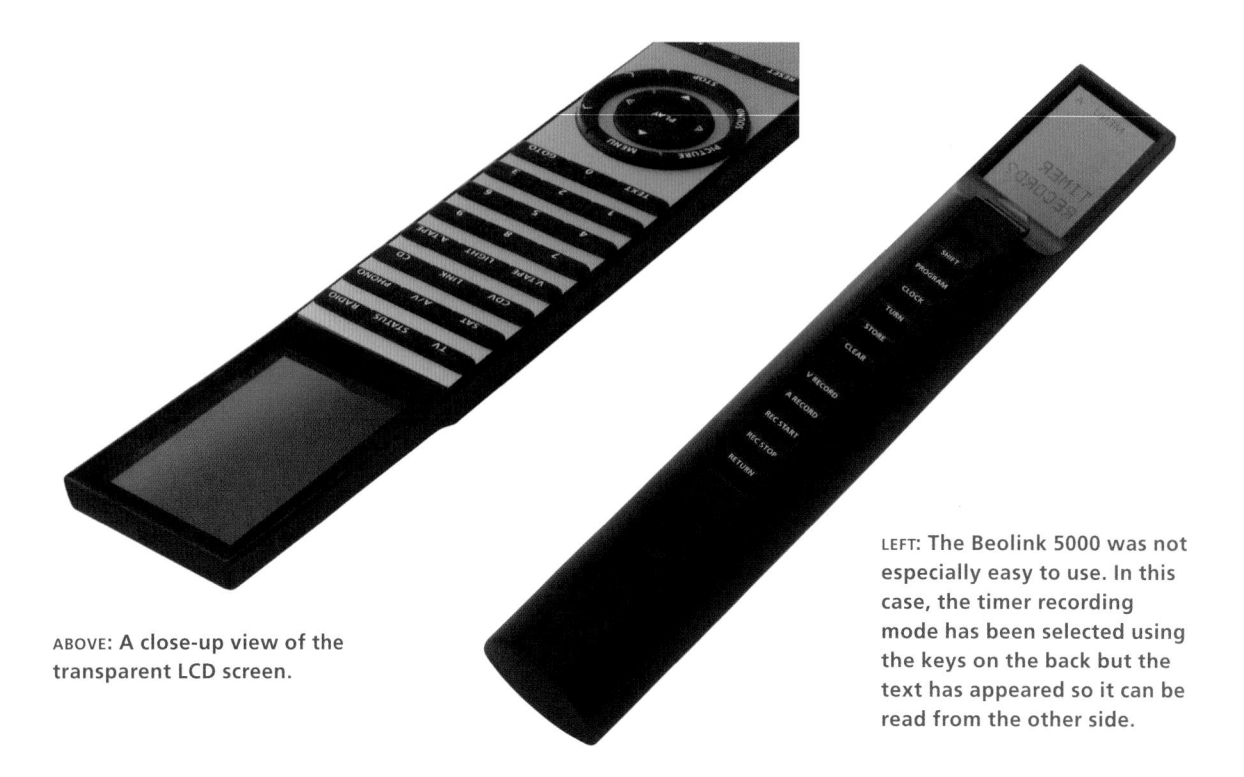

ABOVE: **A close-up view of the transparent LCD screen.**

LEFT: **The Beolink 5000 was not especially easy to use. In this case, the timer recording mode has been selected using the keys on the back but the text has appeared so it can be read from the other side.**

approach had to be taken, so a rechargeable battery was used instead. This was charged either by placing the Beolink 7000 on its supplied table holder or on an optional floor stand, which were both neatly styled but tended to negate the advantage of a cordless remote control.

The Beolink 7000 had one further trick: if the unit was placed flat on a table, touching the aluminium trim at the bottom would cause the upper display part to automatically rise towards the user. This was achieved with a small motor and some gearing. It was made necessary by the poor viewing angle of the LCD panel.

Both these terminals gave access to advanced programming modes within the audio systems, which were not otherwise available; for example, programming the Beocenter 9500 to do different things on different days of the week. 1992 saw the Beosystem 6500 renamed to become the Beosystem 7000, matching the new top of the range terminal. Few other changes were made other than an RDS decoder becoming optional for the radio, which could show the names of FM stations on the Beolink 7000 display.

A Beolink 7000 on its table charging base. The combination of a touch-sensitive LCD screen and two-way infra-red communication made this model amongst the most advanced remote control units ever made.
Availability ●●
Complexity ●●●●

1990 TO 2000: THE SURVIVAL YEARS   151

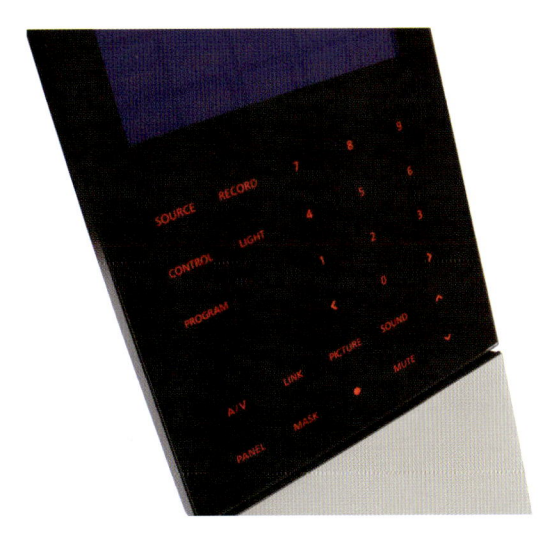

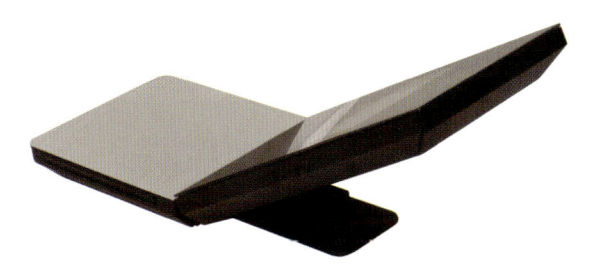

ABOVE: **This view shows the Beolink 7000 tilted up on its motorized base.**

LEFT: **A close-up view of the touch-sensitive keypad.**

When compared to the basic and sturdy Beolink 1000, both these remote control terminals were difficult to use and problematic. The Beolink 5000, in particular, having keys on both sides and a display that could be read from either side, could be very frustrating to operate, as the software allowed only a short amount of time to find the next key to press before the whole operation was cancelled, and the text was often the wrong way round for the position of the key. It was also unreliable, with the battery holder being especially troublesome. The Beolink 7000 was better in many ways but was still far from ideal. With age, the backlight for the LCD can fade away making the display nearly unreadable under normal light – a handicap indeed. The rechargeable cells are also a potential source of trouble and must only be replaced with the correct type. More serious faults are extremely difficult to remedy and, even when new, a factory exchange was recommended if anything went wrong. Both these items are best left to serious collectors only.

Expanding the Range

Both the Beosystem 4500 and the Beocenter 9500 were made in 'budget' versions, the Beosystem 3500 and the Beocenter 8500. As well as removing some of the lesser-used functions, the gleaming, polished finish metal parts were replaced with alternative items

The Beomaster 3500, showing the flat grey metal trim fitted to these models.
Availability ●●●
Complexity ●

1990 TO 2000: THE SURVIVAL YEARS

LEFT: The matching Beocord 3500. This model is technically identical to the Beocord 4500 of the more expensive Beosystem 4500.
Availability ●●●
Complexity ●●

RIGHT: This picture shows the contrast between the two finishes. The cheaper models were painted grey like the example on the left.

painted in a rather uninspiring flat grey. The programmable timers were also not present and the two-way remote control systems were replaced with ones suitable for one-way use only. Even at face value, the saving made by these changes would have been minimal but, on closer inspection, it was clear that only the software programmes had been changed and some of the legends for the touch-panels removed, meaning that production costs were largely unchanged.

Even more bizarre was the Beocenter 8000, a more basic version of the Beocenter 8500 that was introduced (and largely ignored) in 1992. This was similar to the Beocenter 8500 but had connections for basic Beovox loudspeakers only (no Power Link or Speaker Link) and also lacked a microphone input for the cassette recorder. Saving a few pence from the production costs of such a complex and sophisticated unit was clearly pointless, and the public agreed.

The top-line full specification variants of all these models are far more frequently encountered than the basic ones and it is likely that the cabinet finish was the key point over which the original buying decision was made.

Television: The Unity Chassis

The Beovision range of 1989 had fallen somewhat into disarray. Across the models there were three completely different and incompatible chassis: two of which were made by third-party manufacturers. This was clearly not a good idea and considerable savings would be possible if the situation could be simplified. The answer was to design a 'unity' chassis that could be easily fitted to all the models in the range with only minor modifications being necessary; a situation that had existed in the early 1980s but had slowly fallen apart with the introduction of the Beovision MX and LX ranges. The first unity chassis appeared in the LX 4500 and LX 5500 of 1990, which replaced the LX 2502 and LX 2802. It was also adopted in a pair of new MX models, the MX 3500 and the MX 5500, during the following year. The new chassis was designed to drive the Philips 45AX tube system and included such up-to-date features as Power Link, two-way remote control, rotating stand control, the option of a NICAM digital sound decoder (later standard), two SCART AV connectors and an on-screen display. A picture-in-picture module and an integrated satellite receiver could also be installed by a dealer as optional extras. The only model that the unity chassis would not work with was the slow-selling MX 1500 portable, so this was quietly dropped.

The previous generation of sets had been fully software controlled (no knobs to turn and only a few buttons to push) from the user's point of view but for servicing there were still many pre-set controls inside. The new sets were also serviced through the software interface with the picture geometry and colour balancing controls all being set through the one-screen display. Only a few items, such as the tube focus, still required manual adjustment and while this new scheme was generally disliked in the repair trade, it did simplify the initial setting up in the factory greatly.

Although the LX models remained at the top of the range, the new chassis gave the MX models some real credibility at last, for they were fitted with electronics of B&O's own design. The previous MX 3000 and MX 5000, with their Thomson ICC5 chassis, had been

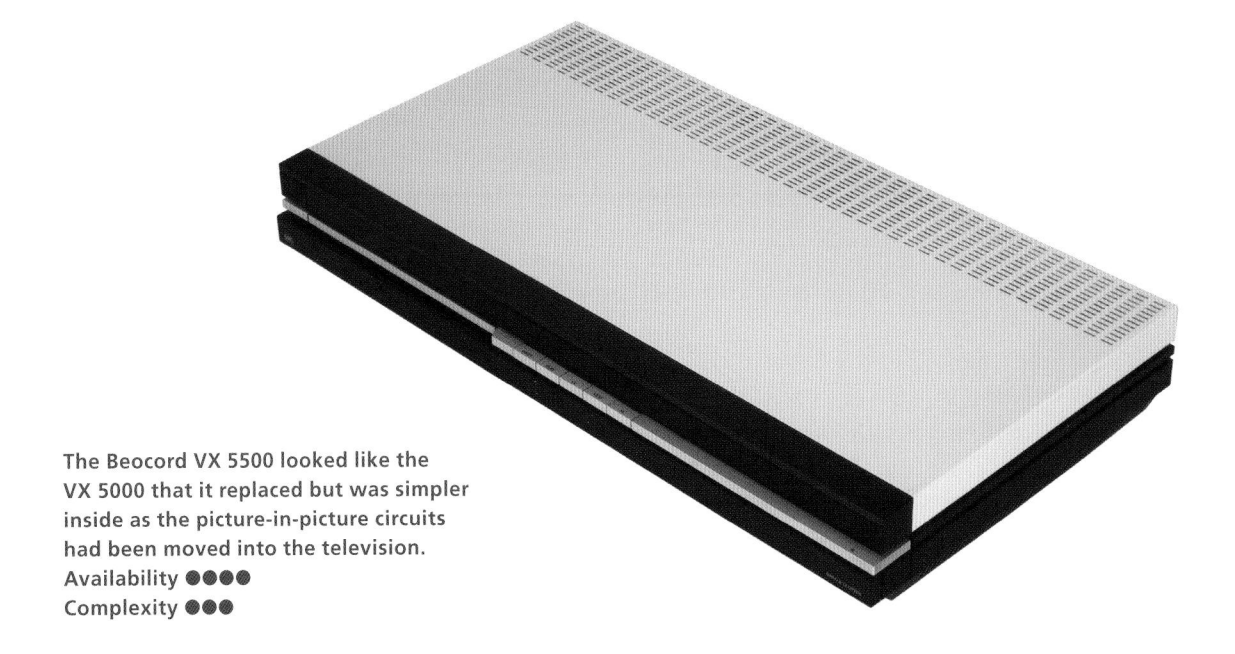

The Beocord VX 5500 looked like the VX 5000 that it replaced but was simpler inside as the picture-in-picture circuits had been moved into the television.
Availability ●●●●
Complexity ●●●

particularly troublesome, capable in the main of fairly mediocre results and it was really only their attractive cabinets that had kept them saleable. With the new models, this problem was at last put convincingly to rest. The new models were, of course, fully prepared for connection to, and integration with, the new audio products (which now mostly featured a special audio connector labelled 'TV') and also with the VX video recorder. Because the television could now be equipped with a picture-in-picture module, the new generation Beocord VX 5500 no longer came with a digital picture processor and consequently lost the excellent tape index function. A manually-entered programme list was fitted instead but as the data was held in the recorder and not on the tape, it was not transferable to other machines.

The VX 5500 is, however, the pick of the range for the collector, as it is simpler than its predecessor and the lower power consumption and less densely packed interior make for better reliability and less daunting repairs. It is also much simpler to operate. The televisions have worn pretty well and the commonality of so many parts means that it is comparatively easy to find a 'donor' set to help repair a treasured example with nasty faults. It can be noted, however, that the printed panels themselves were of rather poor quality in this range. Scorching, discolouration and lifting track-work are all common and so one must be even more gentle than usual when carrying out repairs.

As had been made available previously, a simpler range of TV sets was offered in parallel with the LX series. Now known as the LS range, these sets were largely similar to the LX models but came with simpler loudspeakers and fewer connections for external equipment. Unlike the earlier L models, a teletext decoder was part of the standard package and a rosewood-trimmed cabinet could be specified.

The video recorder range was also simplified during this period. For 1990, only the VX 5000 was available, as the VHS 91.2 and the VHS 82.2 had both been deleted. 1991 saw the previously mentioned VX 5500 introduced along with a new budget model, the V 3000. Intended for use with the LS series Beovisions, the V 3000 was little more than a basic Hitachi model with the words 'Bang & Olufsen' printed on the front. Inside, the software had been changed to work with the Beolink 1000 remote control and, at the same time, the clock display on the front of the machine had been disabled. However, the machine still looked cheap, had an appalling on-screen programming display and even though all the Beovisions of the year were stereo, was only fitted with normal (non hi-fi) mono sound, the last model in the range to be so equipped. The V 3000 was cheaper than the VX 5500 but was still grossly overpriced for what it offered. It would have made far more sense at the time to buy a recorder from elsewhere, if the budget would not stretch to a VX 5500. The only technical point of interest in the V 3000 was that it was the first Beocord video recorder to support two-speed (SP/LP) recording and playback. Previous models from the Beocord VHS 90 onwards had included this feature for sound recordings only.

A Brave New Look

From the point at which B&O became a serious exporter of audio equipment in 1964 until 1991, all the key models shared certain styling cues. The most important of these was the long, low look that so many imitated but no-one else could quite do properly. 1991 saw the introduction of the first major B&O product in a long time that would break with this theme, the Beosystem 2500.

The Beosystem 2500 fitted into the range at a similar point to the Beosystem 3500. It stood upright and in some ways looked like a highly rationalized version of the 'mini' and 'midi' systems that were the mainstay of other manufacturers' ranges at the time, especially when fitted with its matching Beolab 2500 loudspeakers.

The Beosystem 2500 included an AM/FM tuner, a cassette recorder and a compact disc player in one single unit, but it was what the system lacked that made it special. The Beosystem 2500 was designed to address the perceived needs of the age. It was, therefore, designed without consideration for the addition

1990 TO 2000: THE SURVIVAL YEARS   **155**

The Beosystem 2500 brought a fresh new look to the B&O audio line-up. However, despite the striking appearance, most of the technology was conventional.
Availability ●●●●
Complexity ●●

of a turntable, something that gave the designers considerably greater freedom than had been possible with previous generations of equipment. Records are bulky and, in all serious applications, must be played horizontally – a factor that determined the size and presentation of the Beosystem 5000, for example. Once LP records are dismissed, the remaining system can be made slim and upright, just as the Beosystem 2500 was.

The other key realization was that modern solid-state techniques had made powerful amplifiers small and sufficiently cool-running to be mounted in the loudspeaker cabinets. This at a stroke eliminated the amplifiers, along with their attendant heat sinks and mains transformers, from the central piece of the system, making it possible to reduce the size of the cabinetwork without sacrificing power or quality. Moving the amplifiers out of the main unit also freed the stylist's hand, as the larger amplifier components are on the whole inseparable and, therefore, tend to dictate the overall appearance of the design; the Beomaster

The centre unit could be used with any of the Beolab loudspeaker models. The Beolab 6000 was a popular choice.

ABOVE: The styling of the Beocenter 2500 put the spinning CD on show for all to see.

TOP LEFT: As if by magic the glass doors glided open and the CD loader was bathed in a warm glow from concealed lamps at the sides.

BOTTOM LEFT: As the cassette section did not have a door and the heads faced upwards, regular cleaning was required to maintain top performance.

8000 of 1980 is a prime example of this effect. These two factors made the Beosystem 2500 what it was: completely new and an instant hit.

To complete what was already a striking product, the Beosystem 2500 was finished off with its most famous detail, the 'magic doors'. Whilst idle, the front of the system was protected by two glass doors that automatically slid aside when a hand approached. The opening mechanism was carefully designed to give a smooth action and was synchronized with lighting around the control panel, which was faded up automatically at the same time. The magic doors were utterly captivating and must have won the system many sales on their own.

Technically, the Beosystem 2500 was pretty basic. The tuner and cassette sections followed Beosystem

1990 TO 2000: THE SURVIVAL YEARS

4500 practice and offered similar facilities with the exception that the recording level adjustment was automatic and could not be set manually. The cassette mechanism itself used only one motor to operate all the various mechanical functions and, although apparently unsophisticated when compared to the three-motor decks that were found in the other models, this mechanism was eventually adopted across the range. It has performed reasonably well but with age the many plastic parts can wear, often with odd results. It is also unusually intolerant of wear or stretching in the main drive belt, which is particularly difficult to replace.

The compact disc player in the original versions was similar in principle to the Beogram 4500 but in later production a different Philips unit was fitted that used a one-bit digital to analogue converter (DAC). These new units could be recognized easily, as the laser lens moved in a straight line from top to bottom rather than in an arc at the side. One-bit DACs were claimed to be a breakthrough at the time and a real boost to CD quality, but some serious listeners still prefer the earlier sixteen-bit models, especially those based on the Philips TDA1541 chip that B&O used.

Those in search of real innovation could find it aplenty in the loudspeakers. The Beosystem 2500 had only Power Link connectors for the loudspeakers, so only Beolab models could be used, the choice at launch being the Beolab Penta 2, the Beolab 3000 and 5000 or, by far the most popular, the Beolab 2500. The Beolab 2500 matched the Beosystem 2500 in finish, height and the angle at which the front panel was set. Even though they were separate items requiring their own mains supply and signal connections, the two loudspeaker units and the main unit could all be bolted together onto a bracket that concealed all the wiring and allowed the unit to be moved easily.

Inside each loudspeaker enclosure was a 115mm woofer and a 25mm tweeter. The cabinets were fitted with bass-reflex ports at the rear to improve the bass response – a conventional technique. It was the amplifier that was special, for the woofer and the tweeter each had their own power stages, the signal being split early on using a low-level electronic crossover,

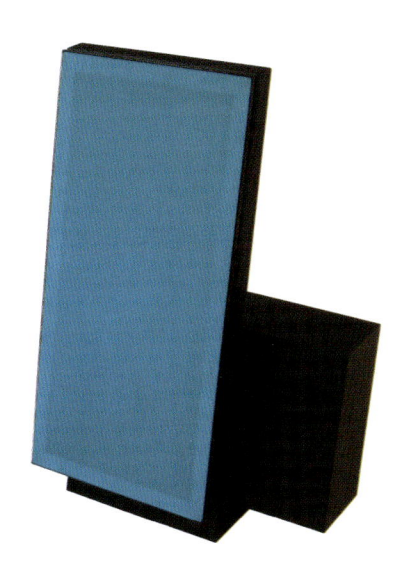

BELOW: The profile of the Beolab 2500 precisely matched that of the Beocenter 2500.

ABOVE: The Beolab 2500 loudspeaker could be used to good effect with other systems but its appearance was awkward because of the housing for the mains transformer.

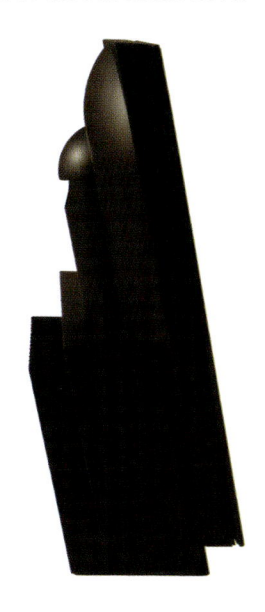

ABOVE: This rear view of the Beolab 2500 shows the bass reflex port at the top and the heat sink for the amplifiers at the bottom.

rather than by passive components connected between the loudspeakers, as was normal practice at the time. The Beolab 2500 loudspeakers could, therefore, be truly described as active, a first for B&O who, in the following years, would design many successful active loudspeakers and become a leader in the field. The active electronic crossovers could be designed with much greater flexibility than was possible with conventional passive arrangements; for the first time, a small loudspeaker could be given a truly 'big' sound. The Beolab 2500 could provide far better bass performance than the Beovox CX 100 and yet was only slightly larger.

There was no reason why the Beolab 2500 could not be used with any model that had a Power Link connection. In practice, however, the shape of the cabinets looked rather odd when not placed with the Beosystem 2500, so they were seldom bought on their own. The coloured cloth covers over the front of the loudspeakers could easily be changed to suit mood or décor, just like the coloured sides of the Beolit radios of the 1970s.

For those on a budget, a version of the Beosystem 2500 was offered without the cassette recorder. This was known as the Beosystem 2300. In lieu of not being able to record anything, the Beosystem 2300 offered extended compact disc programming options, such as track naming. Like many lower cost models, the Beosystem 2300 was comprehensively outsold by the more expensive version. Both were offered in one finish only: black plastic with grey plastic around the control/loading area. The control panel was enlivened by some aluminium trim but this was only the thinnest possible sheet stuck in place by adhesive tape, a type of cabinet finish that had previously been the preserve of cheaper brands. On the whole, all the cabinet materials were uninspiring and felt insubstantial; it was only the excellent work that was done by the stylists that made the appearance of the model so attractive.

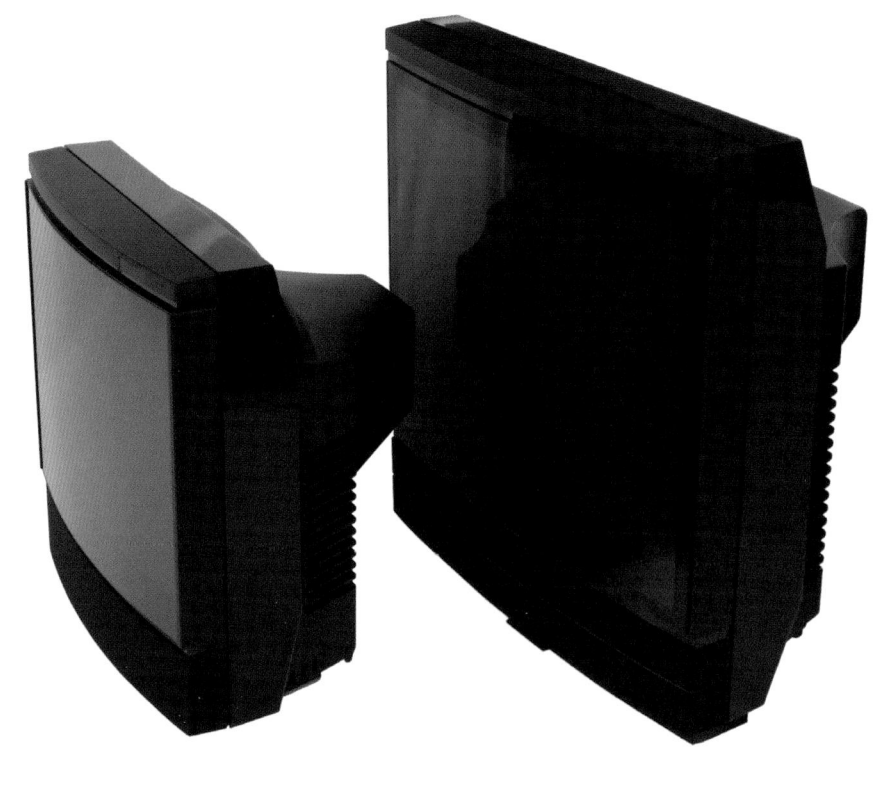

Two best-sellers of the 1990s, the Beovision MX 4000 (small) and MX 7000 (large), were both amongst the best sets to bear the Beovision name. Both sold in huge quantities and deserve a place in any collection.
Availability ●●●●●
Complexity ●●

1992 – Another New Video Generation

The concept of the unity television chassis had been so successful that when its successor was launched in 1992, the original dimensions and basic layout were retained so that the cabinets for the various television models needed little or no modification. The new chassis was slightly simpler in its construction but was still capable of excellent results and good reliability. It would remain in use until the end of the decade and provided the basis for the key mid-range Beovision models that were traditional strong sellers. The MX range remained with a 21 and a 28in model known as the MX 4000 and MX 6000, respectively, whilst the LX range became the LX 5000 and the LX 6000 for 25 and 28in screen sizes. Basic LS models continued to be offered with a reduction of connection possibilities and for the first time a similar set became available in the MX style cabinet, the MS 6000. This came with a 28in tube and a cabinet back in unpainted grey plastic. At the other end of the MX range the 28in MX 7000 of 1993 had active loudspeakers working on the same principle as the Beolab 2500. This model would have a long life in the range and along with the Beovision 7802 was one of the best Beovision designs; a tidy example in good working order would be an asset to any collection. The other great set of this period was the LX 6000. Available until 1996 and with a fine rosewood-inlaid cabinet, this set had a clear lineage that went right back through many landmark models to the Beovision 3000 colour model of the late 1960s. In a world of mass-produced plastic tat, the LX 6000 was the discerning choice and it remains so today.

If even the LS 6000 was too expensive, B&O offered the even cheaper and more basic 28in LE 6000. Rather disappointingly this used a Philips chassis and was housed in a moulded cabinet that looked like an LX set but was quite different in detail. The loudspeakers were also greatly simplified and, in general, the set was not a good one. It could be used with a Beolink 1000 and a Beocord V 3000 video recorder but, if this was as far as the budget stretched, one was better off buying something else. There was also an MX-styled version known as the ME 6000. This was also not really worth buying either and few were sold.

1992 also saw the launch of the final model in the Beocord VX video recorder series. The Beocord VX 7000 looked much like the previous models but internally it had been greatly cost-reduced. Most of the tape transport keys had been removed, so only

No wood inlays or aluminium trim on the Beovision MX range, but there was still great attention to detail in the cabinet design. Although the MX range was available in many distinctive colours, black and silver sets are by far the most common.

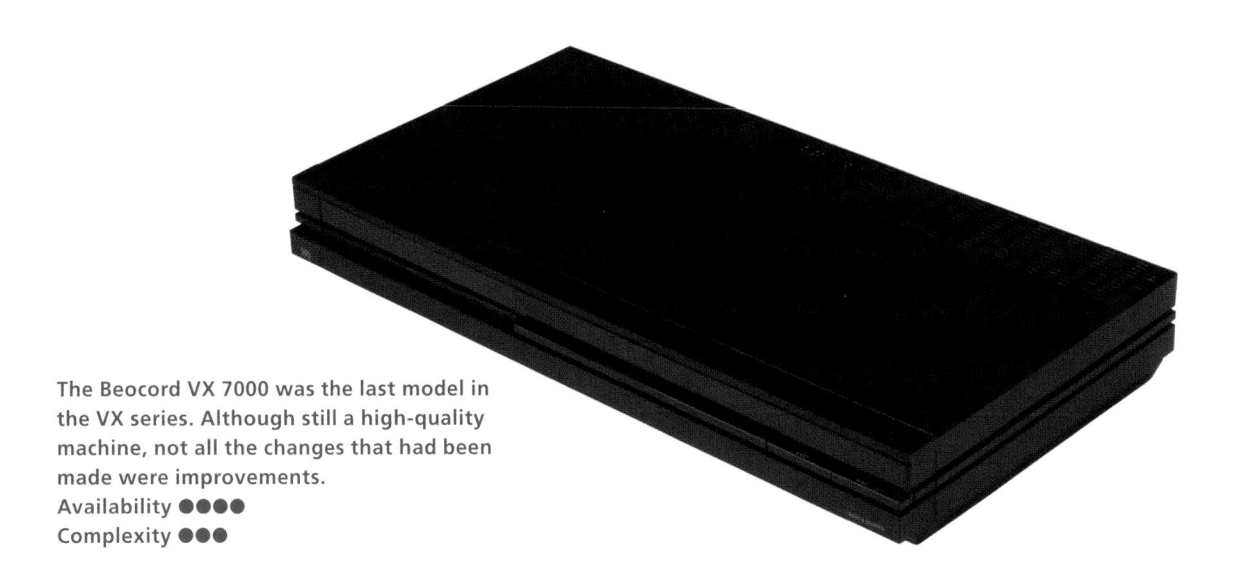

The Beocord VX 7000 was the last model in the VX series. Although still a high-quality machine, not all the changes that had been made were improvements.
Availability ●●●●
Complexity ●●●

remote control operation was possible and as the front panel no longer opened, the audio recording level control was relegated to screwdriver adjustment at the back. The recording level meters were also removed, the only way that remained to determine the correct level was that the standby light now flashed in pause mode when the setting was correct. Even the metal trim at the front had been replaced with a strip of black plastic and the headphone and microphone sockets were deleted. The deck mechanicals were similar to the previous type but even these had been simplified. As a result, the tape remained loaded around the video head drum at all times, even during winding. This needlessly accelerated head wear and could not be seen as an improvement. Only in the 'high speed' rewind mode was the tape fully unlaced, but in this mode the tape counter and index search functions did not operate, a situation that was not ideal. In its favour, the Beocord VX 7000 featured two-speed (SP/LP) recording and a 'flying erase head', which allowed for seamless joins between recordings.

The Beocord V 3000 remained for one more year and was then replaced by the Beocord V 6000. This model was sourced from Grundig in Germany and included hi-fi stereo sound and teletext-controlled timer recording (PDC) for some markets. The mechanicals were sourced from Philips and were known as the 'Turbo Deck' due to the fast rewind time. The mechanism was crude and noisy in action but this did not stop it becoming the mainstay of the B&O video recorder range in the following years. Despite Philips' pioneering work in the field of video cassette recording, these models never matched the VX series in terms of picture quality or slickness of operation. The V 6000 was strangely very large given that there was not a great deal inside it. Again, it represented poor value for money in what had become a very competitive market and was not very popular. Collectors should concentrate on the VX models, unless they are attempting to have an example of every model.

The Home Cinema

Home Cinema is to television what hi-fi is to radio. The aim, as the name suggests, is to re-create the experience of seeing a film in the cinema at home. This first became a practical possibility in the second-half of the 1980s, when stereo video recorders and video disc players were combined with large-screen television sets and quality hi-fi audio systems in the pursuit of the best possible performance. Philips, with their Matchline range and Sony with their Profeel system were amongst the first to offer suitable equipment in

Europe, whilst film studios eventually realized that there was a large amount of money to be made by offering the latest films for rental via high street shops on high-quality video cassettes. B&O equipment was also highly suitable for home cinema use and the company had made great steps towards properly integrating high-quality sound and picture sources.

Home cinema received a great boost when, at the start of the 1990s, surround sound made its appearance. Less than fifteen years after the commercial failure of quadraphonic sound, multi-channel systems were back on the market, mainly thanks to the work of Dolby Laboratories, famous in domestic circles with their noise reduction systems for cassette recorders. Dolby's system was known as Pro Logic and was used to best effect with five separate loudspeakers. The popularity of this system caused many manufacturers to enter the home cinema market with some quite elaborate new models and, in 1992, B&O joined them.

The B&O home cinema system was known as the AV 9000. In its fullest form it comprised a centre unit with a 28in television, a video cassette recorder and a centre channel loudspeaker, all combined into one very large unit, which was mounted on a floor-standing motorized base. In addition, four Power Link active loudspeakers were needed and a modified Beosystem 2500 (known as the Master Panel AV 9000) could also be connected. Finally, a Beolink 5000 remote control terminal was included to complete the system.

Costing as much as a fairly reasonable new car, the AV 9000 was the most complex complete system that B&O had ever offered. In detail, however, most of it had not been too difficult to produce as, for example, the video cassette recorder was a lightly modified VX 7000, the Master Panel was simply a Beosystem 2500 with some parts removed and different programming, and the television used the same basic 'unity' chassis that could be found in the MX 6000. The clever part was the control centre, which was built into the triangular section of the cabinet below the screen. This housed the Dolby surround sound processor and the signal routing stages along with the centre channel loudspeaker and its amplifier. This was

The centrepiece of the B&O AV 9000 home cinema was this elegant unit, which combined a 28in television, a stereo video recorder, a surround sound processor and a centre loudspeaker. All the units were of the highest quality.
Availability ●●
Complexity ●●●●

where the individual parts were brought together as an integrated whole and made the system both easy to use and capable of outstanding results.

To complete the cinema experience, the viewing screen was covered by a pair of curtains behind the contrast screen. Once the picture was ready to be seen, the curtains retracted automatically to reveal it, lending the system an elegance that no other could match. This operation could also be synchronized with automatic dimming of the room lights, if desired. This was undoubtedly home cinema at its finest.

1990 TO 2000: THE SURVIVAL YEARS

Room lighting could be controlled using a B&O light control. This LC2 model was introduced in 1993.
Availability ●●●
Complexity ●

Part of what made the AV 9000 system so expensive was that four active loudspeakers were needed. This remains a problem for collectors today as the loudspeakers have held their value far better than the rest of the system. To acquire it in its entirety is still an expensive (although highly worthwhile) exercise. While it was possible to equip an AV 9000 system with four Beolab Pentas, this pushed the overall price from expensive to ridiculous. Even so, some systems were ordered this way. To make the system more viable, two new types of active loudspeaker were introduced alongside the AV 9000 and, although the latter is long out of production and largely forgotten, the loudspeakers remain in the range, at the time of writing, and are still very popular.

The Beolab 8000 and 6000 were column loudspeakers like the Beolab Penta, only smaller. The Beolab 8000 appeared first and housed within its cylindrical aluminium cabinet two 100mm bass drivers and one 18mm dome tweeter. These were connected to a two-channel amplifier and active crossover that followed the principles set down by the Beolab 2500. The appearance of the cabinets was strikingly elegant and a real departure from the various square wooden boxes that still dominated the scene at the time. Connections were primarily made via Power Link but, as standard RCA-type sockets were included, the Beolab 8000 could be used with other

This tiny dome on top of the AV 9000 monitor houses the infra-red transceiver and the standby indicator. It made a surprising reappearance on the BeoVision 5 plasma television ten years later.

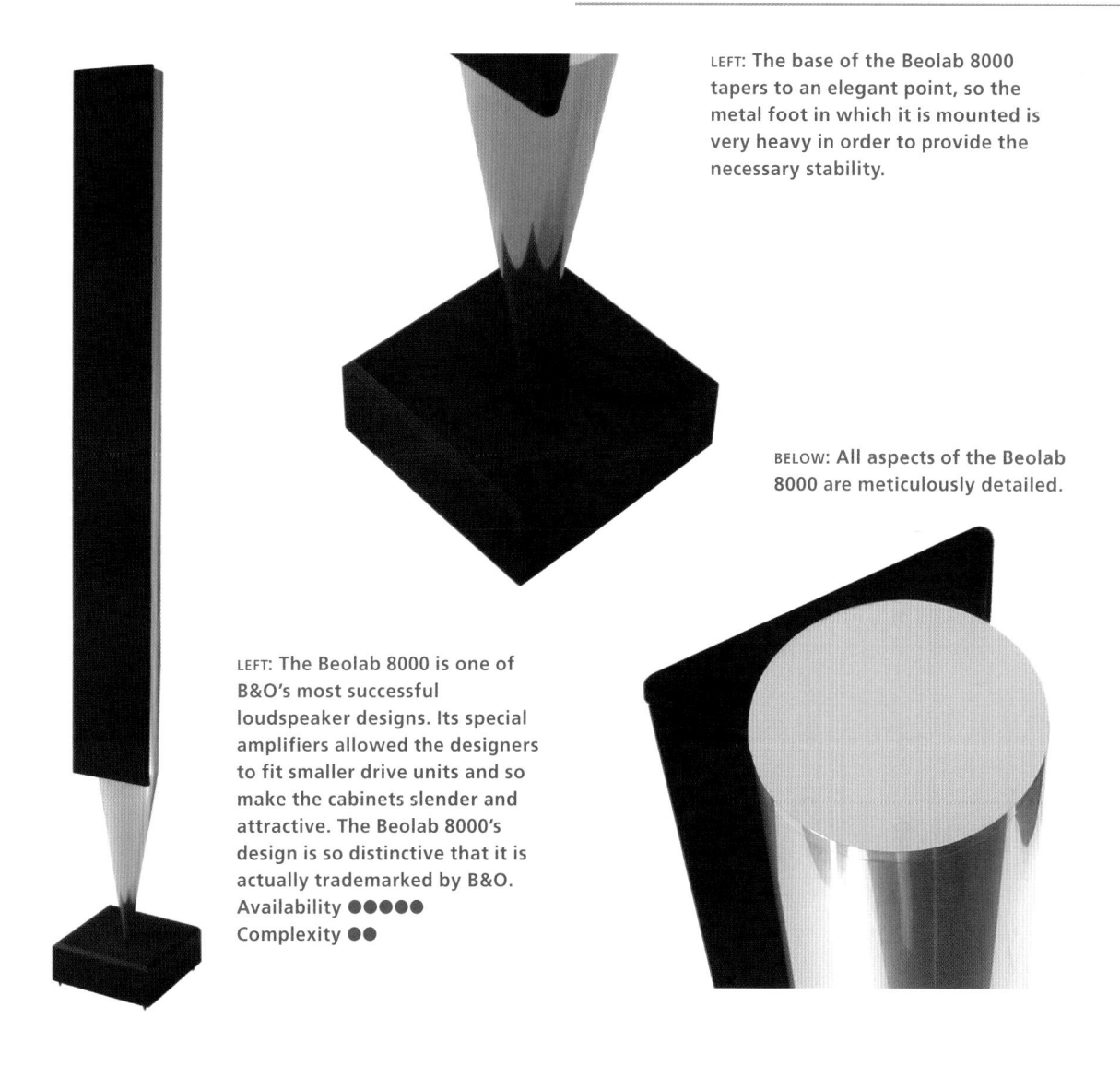

LEFT: The base of the Beolab 8000 tapers to an elegant point, so the metal foot in which it is mounted is very heavy in order to provide the necessary stability.

BELOW: All aspects of the Beolab 8000 are meticulously detailed.

LEFT: The Beolab 8000 is one of B&O's most successful loudspeaker designs. Its special amplifiers allowed the designers to fit smaller drive units and so make the cabinets slender and attractive. The Beolab 8000's design is so distinctive that it is actually trademarked by B&O.
Availability ●●●●●
Complexity ●●

manufacturers' equipment too. This made the mainstream hi-fi press take an interest and, in general, they liked what they heard. Some pronounced the bass to be a little light and, therefore, recommended the Beolab 8000 for classical music only.

The Beolab 6000 followed soon afterwards and was slightly smaller still. It used two 90mm bass drivers and the same 18mm tweeter, along with a slightly less powerful amplifier. In terms of the AV 9000 system, the Beolab 8000 was recommended for the front two channels and the Beolab 6000 for the rear. The loudspeakers soon found useful applications elsewhere in the range too: the Beolab 8000 was the ideal partner for the Beocenter 9500 and the 6000 worked well with the Beocenter 2500 and 2300.

Both these loudspeakers were modified for the following year with a new system called 'adaptive bass linearization'. This was a development of the bass equalizer, which had first appeared in the Beolab 3000 and 5000 panel loudspeakers, and had the same effect of improving the bass performance at low listening levels. Other modifications were needed during the long life of these loudspeakers. The original woofers were not satisfactory, as their plastic cones could become detached from the voice coils if the loudspeaker was played loudly for long periods. The

ABOVE: The circular base of the Beolab 6000 echoed that of the AV 9000 – this would become a theme of many of B&O's floor standing models during the next decade.

BELOW: The detailing at the top of the Beolab 6000 is very similar to that of the Beolab 8000. The fine lines on the aluminium body are not scratches, but artefacts of B&O's aluminium polishing process.

The Beolab 6000 was originally intended for use as a rear loudspeaker for the AV 9000 home cinema but it was also ideal for use with small audio systems.
Availability ●●●●●
Complexity ●●

tweeters were also not sturdy enough and could easily be overloaded; modification packs were included with the replacement tweeters to make the protection circuit more effective. When buying Beolab loudspeakers of this type, it is essential to ensure that all the drivers are working, as it is not unusual to encounter them with defective tweeters that the owner claims not to have noticed.

A New Way to Start

Making 'affordable' products that are not foolishly expensive but still retain the characteristics and quality that one would expect from B&O, was something that the company did with varying degrees of success over the years. One of the more convincing attempts appeared in 1993 with the introduction of the

1990 TO 2000: THE SURVIVAL YEARS 165

BeoSound Century. Priced at just under £1,000 in the UK, this compact model offered a CD player, an FM radio, a cassette recorder and a pair of active loudspeakers in a single, flat cabinet that was designed to hang on a wall or be mounted on an angled stand. Even at this price point, a single 'magic door' was included, helping to give the sense that this was indeed a 'real' B&O design. The introduction of the BeoSound Century coincided with the discontinuation of the Beosystem 4500, ending seventeen years of continuous production for the sleek touch-sensitive range that had started with the Beomaster 1900.

The BeoSound Century was in its original form very, very basic; for example, there was no headphone socket and the radio covered the FM band only, though an AM band was available in some markets. Even the FM tuner could not be used on its own, as the antenna was built into the carrying handle, which

RIGHT: These black buttons control all the functions of the BeoSound Century. Their functions illuminate in red when they can be used.

BELOW RIGHT: In 1993 you could buy a complete B&O music system for 'just' £1,000. The BeoSound Century was an attractive entry to the range.
Availability ●●●●●
Complexity ●

BELOW LEFT: : Slim styling and perfect sound, the BeoSound Century could only have come from B&O.

had to be bought separately. A plastic table stand was included but the wall bracket (a simple piece of folded metal) also cost extra, as did a Beolink 1000, if remote control was needed. There was no programmable timer and the BeoSound Century could not be used as an extra active loudspeaker in a link system, although there was a single connection for an extra sound source (AUX). The cassette section was also very simple: it had automatic recording level only and could not record on metal tapes. The specification suggested that Dolby noise reduction was not fitted either, as the system that was included was simply referred to as 'NR'. However, the Sony processing chips that are used in the Century are in fact Dolby types, so on this point the BeoSound Century was sold a little short.

Despite its shortcomings the BeoSound Century did become very popular and was a far better proposition than the previous Beosystem 10. Later, the AM radio would become standard and a headphone socket was fitted, improving the situation slightly. Like the Beolab 2500, the loudspeaker frets were available in a variety of colours and these could be changed as the owner wished. Collectors should buy the newest and tidiest example they can find.

Widescreen TV

In an attempt to enliven a pretty stagnant market for new television sets in the early 1990s, most manufacturers introduced widescreen sets. In these models, the proportions of the screen were nominally 16:9 instead of the previous standard of 4:3. The development of the widescreen format was rushed and sets appeared before the broadcasters had really agreed on a standard for them and so, as a result, many ended up showing either a small picture with a big black band down each side or a grossly stretched and distorted picture. Expanding the proportions of a picture tube was not as simple a process as one might initially imagine, and because widescreen sets tended to be offered in the larger sizes, there were many technical problems with the early models, something from which even B&O were not immune.

The first B&O widescreen set was the BeoVision Avant. This 28in screen model was a large floor-standing set that included a built-in video cassette recorder. It first appeared in 1994 and sold well straight away. It was positioned in the range just below the AV 9000 and could be fitted with a Dolby Pro-Logic surround sound decoder for home cinema use, if required. Unlike the AV 9000, the BeoVision Avant included a pair of built-in active stereo loudspeakers and so could be used without additional loudspeakers, if stereo sound was all that was required.

The BeoVision Avant brought a new look to television and discreetly brought back the 'console' model. The video recorder loading slot is just beneath the triangle below the screen.
Availability ●●●●●
Complexity ●●●

This digital display bar shows the operating mode, channel number, tape counter or time.

These hidden buttons behind the display give access to the set's basic functions without the need for the Beo 4 remote control.

The video recorder was a Philips design that used the same deck as had been seen in the Beocord V 6000. This unit was of poor quality when compared to the VX-based Control Center VTR that was fitted to the AV 9000. A popular dealer excuse was that the mechanism needed to be 'run in' but what was more likely is that they hoped that the new owners would just get used to it. The television used a new chassis that was specially designed to match Philips widescreen tubes and, as these developed, many changes and modifications made throughout the model's life.

The cabinet designers attempted to hide the bulk and depth of the widescreen tube with an interesting styling idea, the 'wall'. The wall was the glossy painted section that separated the screen from the back of the set. By offering the wall in a range of colours and finishes that contrasted with the back and the screen surround, the large size of the cabinet was effectively hidden by breaking up the shape. The wall was constructed from fibreboard and finished in a car-type metallic paint, which made the set look really special. The material was quite soft, however, and the corners did dent rather easily. Performance was initially quite good but the positioning of the on-screen menus in the corners of the screen did show up that, in some cases the tube convergence was poor. Deliveries were suspended early on in the life of the model due to tube problems, but when they resumed, the sets sold in huge numbers.

The BeoVision Avant came with a new type of remote control terminal, the Beo 4. This terminal fitted into the range between the Beolink 1000 and the Beolink 5000, offering the functionality of the former with the appearance of the latter. The Beo 4 appeared to have a large LCD screen like that of the Beolink 5000, but in fact it could only display a single line of characters in the middle. It was also not a two-way design and so the screen could only show the status of the terminal, not the equipment it was operating. The idea of the Beo 4 was to simplify the operation of all types of equipment by moving the various menu choices onto the remote control itself. A further feature was that the available functions were called up from a list that the user could customize, keeping the button-count reasonable and hiding those functions that were not relevant to the system the owner had bought. The Beo 4 was instantly recognizable by the four coloured keys in the centre of the main keypad. The colours were familiar from the 'fastext' teletext menus that most television sets of the period offered but, in the case of the Beo 4, they operated additional functions on other equipment too.

The Beo 4 unfortunately signalled the end for the two-way infra-red remote control system. In the 1995

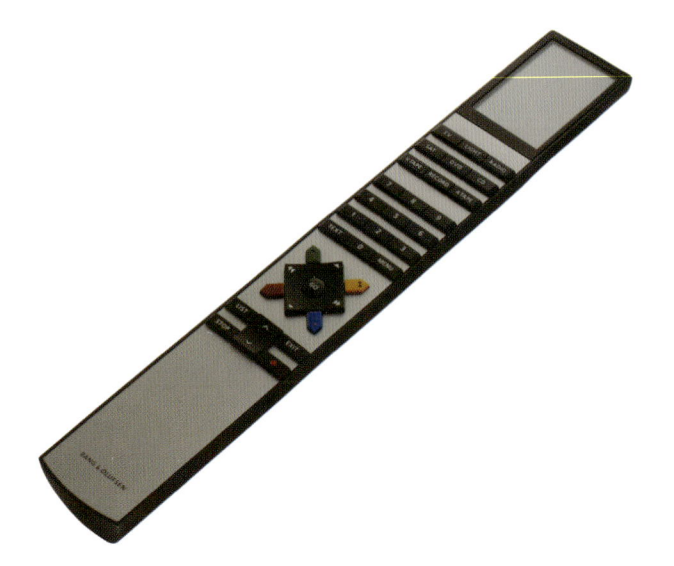

ABOVE: **The Beo 4 replaced the Beolink 1000 as B&O's universal remote control.**
Availability ●●●●●
Complexity ●●

RIGHT: **Although the Beo 4 replaced the Beolink 1000, the influence of the Beolink 5000 (far right) on its design is clear.**

range, all the two-way models had either been discontinued or replaced with simpler one-way alternatives. Some of these can easily be recognized as the word 'Beo4' was simply added to the end of the model names, e.g. MX 7000 Beo4. Modifications included redesigning the menu system to suit the way the Beo 4 operated and the removal of the infra-red transmitter that had been the key part of the two-way system. Other models were simply downgraded; for example, the LX 6000, which used to come with a Beolink 5000, now was supplied with a Beolink 1000. The VX 7000 video recorder was made suitable for both systems with a setting that could be altered in the service menu. The Beo 4 was, of course, compatible with earlier models as well, but it could be difficult to use for certain advanced functions. The two-way system had undoubtedly been the most advanced TV/audio remote control arrangement available anywhere but in its final full implementation it had lasted only five years.

In 1998, a larger screen version of the Avant was introduced. This effectively replaced the AV 9000 system, which had remained in the range as a slow seller up until that point. The 32in Avant included 100Hz scanning, a method of reducing the effect of flicker in the picture. While this undoubtedly worked, there was a lot of digital processing involved and this did have the effect of reducing the picture detail to some extent, which was unfortunate. All Avant 32in sets included a Dolby Pro Logic decoder as standard but, as the built-in stereo loudspeakers were retained, the owner was not compelled to buy the four extra Beolab loudspeakers that would be required to use it, if they didn't want to.

A collector should be wary and selective if they are considering buying a BeoVision Avant. The picture tubes that were used were of very poor quality and failure is many times more likely than with any other Beovision television set, even models that are much, much older. Any set that does not display a perfect picture should be treated with extreme suspicion and it is wise to assume the worst as the chassis has proved to be very reliable.

To be worthy of a place in a collection, the set should be in full working order and that includes the video recorder. These units were not that durable and the slot is at just the right height for children to post things into, something that can damage the flimsy plastic mechanism quite badly. Many of the video recorder sections have fallen into disrepair and owners have simply connected external units instead of paying for expensive repairs. Other machines based on the Philips Turbo Deck can be used as a source of parts but bear in mind that the video head drum cannot be replaced without the use of special tools. Finally, inspect the cabinet carefully. Such a big set is difficult to handle, so the cabinet edges and corners will soon get scruffy if the set has been moved about a lot or has been stored carelessly. Always remember that many BeoVision Avants were sold, so there is no need to settle for anything less than a perfect example; keep your fingers crossed that the tube holds out.

A Lull in High Fidelity

1994 was also the last year that the Beosystem 7000 was featured in the catalogue. Despite being heavily cost-reduced in some areas, when compared to the original Beosystem 5000, the 7000 was still an attractive suite of equipment and, set up carefully, was capable of very pleasing results. Even though none of the components could be described as the 'best ever' when compared to the highlights of the previous ranges, it was still a proper separates hi-fi set, B&O's last in fact.

A quieter change had occurred in the Beocenter range. The Beocenter 9500 and 8500 had been dropped and replaced by the BeoCenter 9300. Outwardly the changes were so minimal that they could only be seen once the set was in use; inwardly the differences could not be more marked. The BeoCenter 9300 was the result of extreme rationalization and so, instead of being a complex product constructed from specially made assemblies, it was heavily based around the Beocenter 2500. In as many ways as were practical, the printed circuit units from the 2500 were simply fitted into the cabinet of the 9500 to form the new model. This, of course, led to a drastic reduction in the facilities, with the tape deck being the most heavily affected part. Features such as the Dolby B/C processor with automatic selection, automatic or manual level control and the microphone amplifier, were dropped at a stroke, as was the phono pre-amplifier stage and the two-way infra-red communication system. It has to be remembered that the Beocenter 2500 was a considerably less sophisticated unit than the Beocenter 9500 and cost quite a lot less, so these changes represented a down-grading of the top Beocenter model.

At the same time, the Beocenter 2500 was renamed to the BeoSound Ouverture, while the Beocenter 2300 remained unchanged. The Ouverture received a new type of connection, the Master Link (ML) connector, when the name change occurred. The ML connection was used for multi-room setups and AV integration, and became a feature of all the new models. The BeoSound Ouverture with its ML connector made the Master Panel AV 9000 unnecessary, so this model was dropped at this point.

The introduction of the prefix 'BeoSound' for the Century and Ouverture marked the end for audio products with separate identities. The buying public were now less educated and technically aware, making terms like 'Beomaster' and 'Beogram' too complex, 'BeoSound' and 'BeoVision' were clearly the limit of what could be easily understood. At the same time, the first letter after the 'Beo-' prefix (the 'S' in BeoSound, for example) was capitalized in all product names. While typographically unattractive, this was becoming the fashion in all industries and B&O had begun adopting it two years earlier with the 'BeoSound' name.

These model changes made 1995's audio line-up seem very sparse indeed. There was still a range of fine loudspeakers, from the Beolab Penta 3 (similar to the Beolab Penta 2 but with a slightly different style of display) at the top of the range, through the Beolab 8000 and 6000 columns and the Beolab 4500 panel loudspeaker (a revised form of the Beolab 3000), down to the Beolab 2500. However, the only audio systems that they could be used with that remained were the BeoSound Ouverture, the Beocenter 2300 and the BeoCenter 9300, all essentially the same thing and all based on a model that, when introduced, was effectively at the entry level.

This situation reflected the conditions of the market at the time. Home cinema equipment and what were referred to at the time as 'multimedia PC' computers were taking sales away from hi-fi and it was at this stage that the market split into two sectors: one for small music systems, for those with an interest in reasonable quality but no time for technical issues; and one for the truly bizarre and esoteric, where cost and common sense were largely irrelevant. The previous middle ground, which B&O had shared with the upper end of the mainstream Japanese maker's ranges and the more sensible end of British production, was ceasing to exist. There was no point in introducing new products for a market that was not there. Product areas such as turntables and cassette recorders, where B&O either held key patents or enjoyed a particular mastery of the technology, were coming to the end; these were difficult times.

Telephone Update

Mention has already been made of the Beocom 2000, the first widely distributed Bang & Olufsen telephone. Various models followed, all neatly styled and tastefully coloured. The Beocom 2000 itself lost its coloured keys and bright plastic finishes in favour of more sombre tones; collectors should hold out for the more distinctive early models.

Attempts were even made to re-brand Ericsson mobile telephones, resulting in the Beocom 9000 and

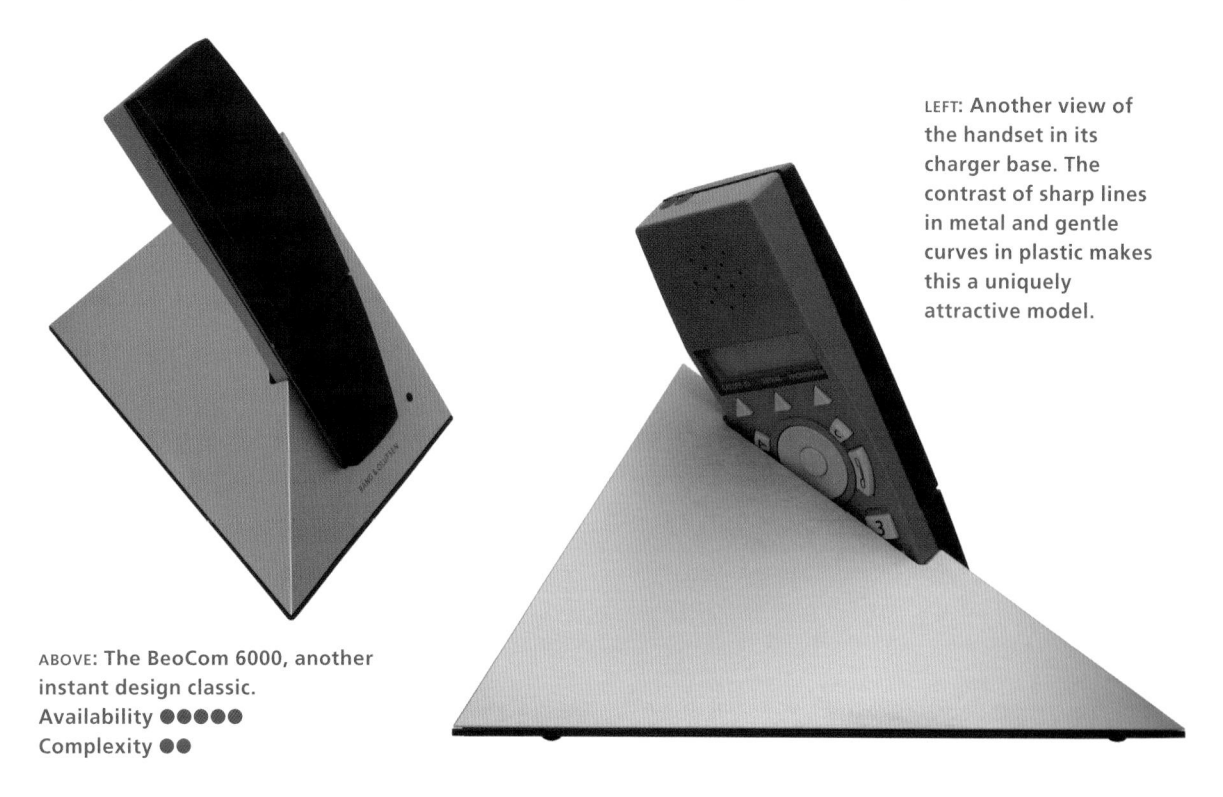

ABOVE: **The BeoCom 6000, another instant design classic.**
Availability ●●●●●
Complexity ●●

LEFT: **Another view of the handset in its charger base. The contrast of sharp lines in metal and gentle curves in plastic makes this a uniquely attractive model.**

1990 TO 2000: THE SURVIVAL YEARS   **171**

RIGHT: The BeoCom 2400 was another successful 1990s Beocom design.

ABOVE: Who else but B&O could give the humble domestic telephone such a beautiful sculptural form?

BELOW: The simple and logical BeoCom 6000 keypad. The wheel in the centre controls many of the telephone functions, as well as being a volume control for B&O audio and video systems.

BELOW: The BeoCom 2400 keypad had an interesting curved shape.

9500. These were not a success; the 9000, in particular, looked like a blue plastic child's toy. The nicest thing to come out of the whole project was the desk charger for the 9500, which was both practical and elegant.

Domestic cordless telephones were also offered. The first model, which appeared in 1995, was the Beocom 5000, a re-branded and very lightly made-over Ascom design that was rather poorly styled. The casework was distinctly 'fat', giving the impression that it was struggling to contain what was inside. The handset had no display but there was a notepad, on which frequently used numbers could be written, that popped up out of the base unit when a button was pushed. The most interesting variant of the Beocom 5000 included a Beolink 1000 compatible remote control function that could adjust the volume of audio and video sources.

An altogether more successful cordless model appeared in 1998. Called the BeoCom 6000, this was a strikingly styled piece of equipment, with an asymmetric handset rising from a metal pyramid base. The BeoCom 6000 used the new DECT digital standard as its basis and gave a level of call quality that was on a par with that which could be obtained from a quality corded instrument. The DECT system allowed the BeoCom 6000 base to be used as a home digital exchange, which could support six handsets between which toll-free internal calls could be made. Operation was simple, as the main control was a large thumbwheel in the centre of the keypad in conjunction with a two line LCD screen.

Both the Beocom 2000 and the BeoCom 6000 make excellent and practical collector's pieces. Both look best in bold colours but these are unusual because they were not the most popular choices when new.

Re-inventing Compact Disc

Having settled for an audio range based around combination units, what was required was an exciting new model. This would burst onto the scene in some style in 1996 as the BeoSound 9000.

At a time when most compact disc players were square black boxes where the disc was loaded into a drawer, the BeoSound 9000 looked like something from another world. The basic idea was to provide a CD changer that could take six discs at a time and play any track from any one instantly. Compact disc changers had been a feature of other makers' ranges for some time but they tended to be styled just like a conventional single-play machine and take an age to change the disc, a time during which various nasty clanking and grinding noises were frequently heard. The BeoSound 9000 put all the discs on show and made the player move around them. This made the disc changes swift and practically silent, thanks to the

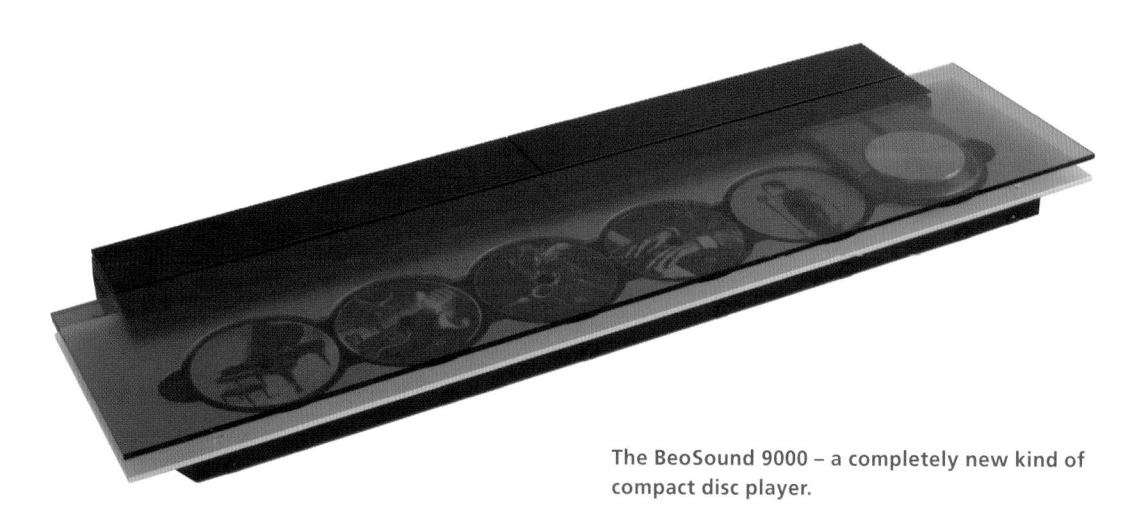

The BeoSound 9000 – a completely new kind of compact disc player.

1990 TO 2000: THE SURVIVAL YEARS   **173**

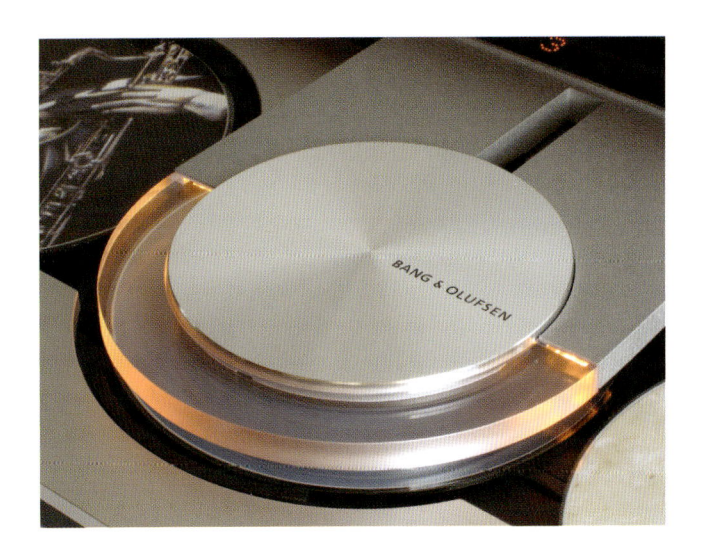

TOP LEFT: **A close-up view of the disc clamper. The clear strip around the edge illuminated when a disc was playing.**

TOP RIGHT: **Discs were clipped into a circular bay, ready to be picked up automatically by the disc clamper.**

BELOW RIGHT: **This is not a large remote control unit! The control panel could be reversed to suit how the system was mounted.**

excellent mechanical engineering of the highest precision that was required to make it work properly. The discs were protected by a glass lid and, although the machine could be used with the lid open, the disc clamper moved at a far more leisurely pace under these conditions to protect the user's fingers. Optical sensors stopped the mechanism completely if there was something in the way and also ensured that each disc was returned to the same angular position as it was in when it was first picked up. If the discs were initially placed neatly, they would remain like that even after they had been played many times. An AM/FM radio tuner was also built in but this could only be used if Power Link loudspeakers were connected. If they were not, then the software assumed that the system was being used as a Compact Disc player in a larger setup and disabled all the unnecessary functions.

Various placement options were offered for the BeoSound 9000 based around a range of stands and brackets, allowing the system to be oriented horizontally or vertically. The only changes that were required to switch the unit between horizontal and vertical orientations were to re-set the suspension springs through a pair of small holes in the rear and to lift out and turn round the control panel, an action that would cleverly also reverse the text display that was built into the clamper. Even though it was a much older design, the Beolab 8000 became the natural loudspeaker to partner the BeoSound 9000, as the system was displayed at its best when the BeoSound was mounted vertically on its floor stand.

Despite being, generally, a very well-made product, the BeoSound 9000 has two significant weaknesses that the collector should bear in mind if they plan to buy one. Firstly the glass door of the early models could come un-stuck from the hinges. Improved doors were made available, which not only used a better adhesive but also had check straps that would restrain the glass should the bonds fail again. Be very wary of attempted repairs with glue in this area by previous owners; a new door of the improved type is the only acceptable method of repair.

The other problem area is the laser pickup. The Philips type that was employed is not particularly durable and this, combined with the unit's likely role of providing continuous background music, means that laser pickup failure is quite a common problem. A number of different types have been employed over the years and they are all expensive and very tricky to replace. Make sure the unit can play any disc quickly and easily, and that it is not excessively mechanically noisy.

Another Format Gaffe

The 1990s were a time of change in the consumer electronics industry. Having recovered from the expensive debacle that had been the video format wars of the previous decade, it soon became clear that no lessons had been learned, as yet again rival formats slugged it out in the market place. Unfortunately, this led to established products, such as the audio compact cassette, being marginalized before a replacement, which the public agreed to be acceptable, was established. Digital Audio Tape (DAT), Digital compact cassette (DCC) and MiniDisc, all absorbed fortunes in development and marketing costs and all would be forgotten within a short time. B&O did well to keep out of this particular battle, never offering machines in any of these formats. They did make one big mistake during this period, however.

The dominance of the compact disc as the pre-eminent pre-recorded audio format turned interests towards the possibility of storing pictures in a similar

The BeoCenter AV5 with its loudspeakers extended ready for use. This attractive model was a poor seller because the CD-i format was not the success that Philips had hoped it would be.
Availability ●●●
Complexity ●●

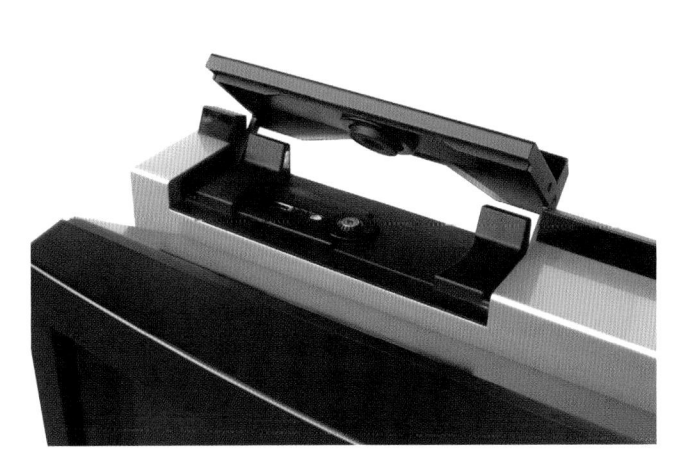

The disc was loaded here and its exposed edges could be seen rotating during play.

The motorized loudspeaker enclosures extended when the set was turned on. At the same time, the motorized stand would turn the whole set to the user's pre-defined viewing position.

manner. Forgetting that just over ten years previously, video disc players had been un-saleable dust-gatherers on dealers' shelves, the industry willingly poured the necessary funds into the development of yet more video disc systems. Philips predictably led the European assault with a format called CD-i, short for compact disc interactive. This initially appeared as a games format to which interest was added with the promise of a 'full-motion video adaptor'. Although this did eventually appear, technical difficulties meant that there had been long delays, by which time the Japanese had struck back with their digital video disc, which when renamed as the digital versatile disc became the now ubiquitous DVD. B&O, however, chose to go with Philips but rather than just putting out a black-box set-top player, they produced a whole new television with integrated CD-i to showcase the new format.

Known as the BeoCenter AV5, and launched in 1996, the B&O CD-i television was the most expensive CD-i machine to be offered to the public. It looked like a shrunken BeoVision Avant and was fitted with a 25in 4:3 picture tube. As well as the CD-i player, the BeoCenter AV5 included an FM radio and a novel loudspeaker arrangement, where two small enclo-sures appeared automatically from the sides of the cabinet when the set was in use. The sound of these was bolstered by a sub-woofer at the rear, the whole ensemble being capable of very pleasant results for a one-piece system.

Of course, as we now know, CD-i was not a success, something that was already becoming clear when the BeoCenter AV5 was launched. Considerably more expensive than the larger BeoVision Avant, which also included a useful video recorder in the package, it was obvious that the BeoCenter AV5 would fail, which of course it did. The bold cabinet colours probably did not help either and some dealers were saddled with metallic orange examples that proved very tricky to shift. Realizing they had made a big mistake, B&O later removed the CD-i equipment from the BeoCenter AV5, leaving it capable of playing audio compact discs only, and slashed the price. However, it was still expensive and lacked both purpose and market appeal.

Some collectors delight in a brave failure and for these people a BeoCenter AV5 is a must. Other than the CD-i sections, the design is fairly conventional and follows BeoVision Avant practice, so similar guidelines apply when inspecting. The picture tube is slightly less troublesome than the Avant type but, as the set is several orders of magnitude rarer, one cannot afford to be so picky if it looks dubious. It is easier to find another tube than another set. With the CD-i section, you are very much on your own. Even the full service manual offers very little guidance in this area beyond a diagnosis flowchart and the special integrated circuits in the decoder have long since become unobtainable.

The BeoCenter AV5 was originally supplied with a special B&O demonstration CD-i disc, which even contains some interactive content. It is well worth asking the owner if they still have this, as it is quite interesting and may well be the only CD-i disc you ever own!

Goodbye to Some Old Friends

As the 1990s drew to a close, more long-running models, which had slowly gone from strong sellers to

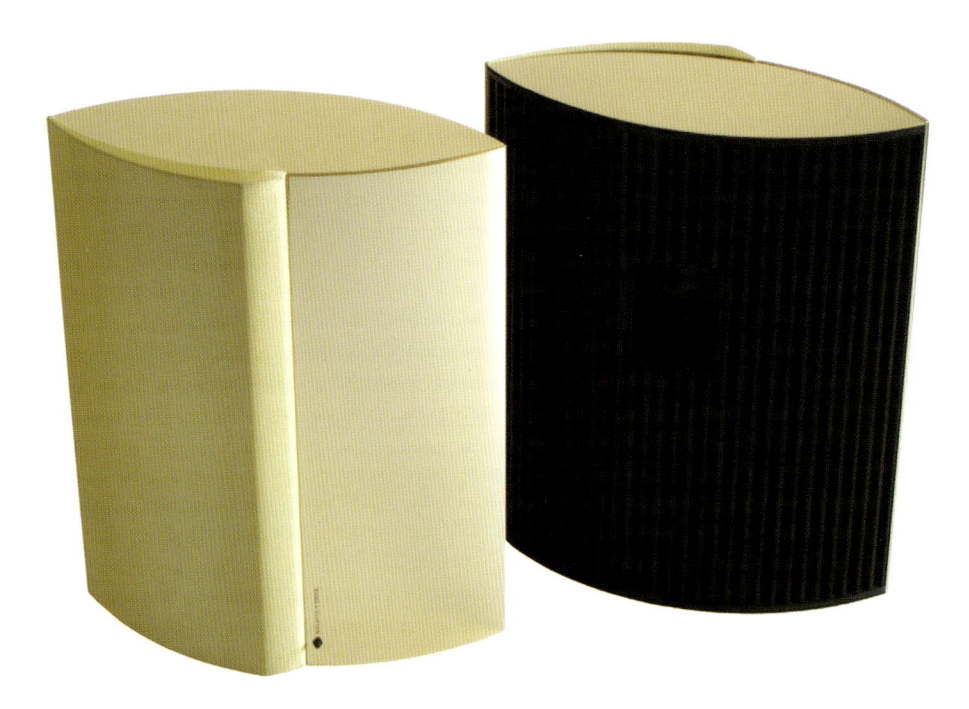

ABOVE: **Whilst gold is not a traditional B&O finish, it cannot be denied that it suits these BeoLab 4000 loudspeakers perfectly. The large heat sink on the rear keeps the internal amplifiers cool.**
Availability ●●●●●
Complexity ●●

RIGHT: **An unusual feature of the BeoLab 4000 loudspeaker is a power switch on the front. The small round button also houses the standby indicator.**

curious backwaters in the range, were quietly discontinued. 1997 saw the end of the Beovox RL 6000, the last big conventional passive loudspeaker in the range, along with the Beolab 4500, which was beginning to look very dated alongside the sleek new offerings. The removal of the Beovox RL 6000 from the catalogue meant that there were no longer any full-sized loudspeakers of conventional design for buyers of the BeoCenter 9300 (the last audio product in the range capable of driving them). The only Beovox models that remained were the CX 50 and CX 100, which had applications in multi-room systems, although these too would be discontinued in 1999. Next to go in 1998 were the Form 1 headphones (arguably the most practical of all the various designs that had been offered over the years) and the Beocord VX 7000 video cassette recorder, along with the cheaper V 6000. Both of these were replaced by the BeoCord V 8000, a rather half-hearted offering that was essentially the video recorder section of the BeoVision Avant distilled into a small free-standing box. Styled along the same lines as the Beocord V 6000 but without the bulk, this would be B&O's final VHS machine.

As if they were concentrating on things they were good at, B&O re-stocked their range with yet more active loudspeakers. An attractive new model, the BeoLab 4000, appeared in 1997. Technically very similar to the Beolab 2500, this new design featured a polished metal cabinet anodized in some truly striking colours. In profile, the cabinet took the shape of a leaf, a welcome break from the traditional conformity that typically afflicted loudspeaker design.

Another key model appeared in 1999. The BeoLab 1 was a large column model that was placed in the range above the Beolab Penta 3. The BeoLab 1 was larger than the Penta but contained only four drive units (in comparison to the nine in the Penta 3), although it was a truly active design. The amplifier for the bass drivers used a new B&O technology that was given the name 'ICEpower'. Strongly hyped at the time, ICEpower was little more than a re-invention of the class D/PWM amplifier, which had cropped up on a number of previous occasions, most notably by Sony who had some success with it in the late 1970s.

The ICEpower amplifier enjoyed the usual class D benefits of high power output without the need for large heat sinks. This allowed the BeoLab 1 to have sleeker lines than were possible with the Beolab Penta, whose conventional amplifier needed generous ventilation if worked hard.

Close of Play Scores

The coming of the 2000 season marks the end of our B&O story. The company had done well to survive the financially difficult 1990s and emerged at the end as still both the designer and the manufacturer of most of their range. The market for all entertainment equipment had changed markedly in the preceding ten years; quality high-fidelity equipment was now a niche interest and had largely disappeared from high street shops, although new subscription services had maintained the public interest in television. The B&O range of 2000 contained four ranges of televisions, five integrated music systems, six types of active loudspeaker and four models of telephone. In addition, a small range of products remained for use with the multi-room system, one of the few pieces of unique B&O technology that had survived the culls.

There was little that was new in the 2000 catalogue, but one product did point the way to the future. The BeoVision 1 was a 25in television that was housed in a distinctive coloured plastic cabinet that could be mounted on a variety of bases and stands. The screen size placed it between the small and the large Beovision MX models but a few extras, such as active loudspeakers and a motor for the stand, were included as standard. The set came with a new type of remote control, the Beo 1. This was an unusual design, executed in polished metal. Despite being interesting to look at, the Beo 1 was functionally poor as there were not really enough keys to operate all the functions of the television quickly and easily, and the shiny casing showed every finger mark. It was soon dropped and a Beo 4 was included with the sets instead.

What made the BeoVision 1 new, however, was not how it looked or how it worked, it was how it was maintained. Instead of publishing a circuit diagram and a parts list, the service backup consisted of a complete chassis exchange scheme available only through B&O franchised dealers. This arrangement was easier for the manufacturer to accommodate but unfortunately made every repair an expensive one and would leave the sets unsupportable once the factory backup ceased. Collectors should be careful as the situation is far from ideal in terms of long-term ownership or restoration. Similar schemes would eventually become the norm across the whole range with only a small selection of replacement parts and assemblies being offered for each model. This has brought about the bizarre situation of having to scrap nearly-new equipment because the parts to repair it are not available.

The BeoVision 1 was also amongst the first B&O products to require a password to be entered to resume operation after the power had been disconnected. Again, collectors should be careful.

ABOVE: **Basic VTR controls were hidden at the rear.**

LEFT: **The Beo 1 looked pretty but poorly thought-out controls and material choices meant that it was not a great success.**

CHAPTER SIX

PRACTICAL COLLECTING

Now that the important B&O models from the collecting period of interest have been put into their correct historical context, it is appropriate to move on to the skills and knowledge that a successful collector will need. The following seven sections go through the basics and can be regarded as a grounding on which further study can be based. Nevertheless, there is sufficient information for a beginner to start collecting without too much fear of mistakes or accidents. The first section deals with safety, not the most glamorous or interesting of subjects but still an important one. Resist the temptation to skip over it, as collecting is no fun if you hurt yourself doing it.

Section One: Safety Issues

Before handling, cleaning, adjusting or maintaining any of the equipment in your collection, it is important to recognize the risk factors that are involved. B&O equipment of the period under discussion is in general safely constructed to standards similar to those in force today, but there are still aspects that should be approached with caution.

The first rule of workshop safety is only to do what you fully understand. Do not guess or just 'hope for the best' in any situation or you may be putting yourself (or others) in danger.

Be careful when lifting equipment. Some B&O models are both very heavy and awkwardly shaped for a person working on their own to lift; so if in doubt, get help. Television sets are the main problem area, but some audio models are also very heavy and, in addition, have sharp metal edges in unseen places.

As with all electrical equipment, the main risk you will face while working on things from your collection is electric shock. Whilst not always fatal, serious injuries can result from accidental contact with live parts, such as deep cuts from metal edges and the tips of screws, which result when a hand is quickly jerked away. You may also jump back or fall over, so make sure the work area is uncluttered. Much has been written on the effects of electric shock and the first-aid measures needed when it occurs; read and understand the basic rules and avoid working alone when possible, if you think there may be a risk. Always disconnect the power cable of anything you are cleaning, inspecting or dismantling before you start work.

Another danger exists in the form of the high vacuum picture tubes used in televisions. If ruptured suddenly, the glass envelope can fragment violently, showering the work area in shards of glass. This was a favourite subject in older repair manuals but the generation of sets covered in this book have been designed to reduce the risks considerably. Even so, work with care around the tube. Don't knock or stress any part of it and never attempt to lift it out of the cabinet by the neck. Modern tubes have been made safer by the use of armoured glass and the fitting of a 'rimband' around the perimeter of the screen. The screen can be weakened by very deep scratches and the rimband is only effective if it is tightly secured, so work very carefully around any tube with defects in these areas. Dispose of worn out and unwanted tubes carefully, as they contain toxic chemicals.

Many B&O models featured in the book have electronic timers that can be programmed to turn the equipment on at any time. This can cause a hazard if an item turns on unexpectedly whilst being inspected or when it is unattended. Don't confuse the 'standby' mode with the security of physically removing the mains plug from the wall socket before starting any work or leaving the work area for any reason.

Battery operated models are generally safer to work on but care is still needed. Modern batteries (especially rechargeable types) can supply a very large amount of current if short-circuited, enough to cause a burn or start a fire. Models that include microcomputer control often have small batteries to maintain the computer memory. Whilst these are generally safe, some types contain lithium and may be subject to local laws concerning their safe disposal. Consider this whether you are disposing of either just the battery or a whole piece of equipment. Local government agencies will be able to inform you of the rules covering the disposal of batteries.

Certain parts of some electronic equipment are carefully specified to ensure safe operation. Items such as fuses, mains switches and power transformers are obvious examples but there are many more whose contribution to the overall safety of the original design is rather more subtle. Things such as isolated aerial sockets, rotary controls with insulted plastic shafts, flame-proof 'fusible' resistors and plastic screws are all carefully chosen during manufacture, so do not substitute them for other parts that may leave the user unprotected. This is especially important if you plan to sell the equipment, give it away or allow inexperienced people to use it. Always re-fit any protective covers or insulated sheets that may have been removed to gain better access to a part or assembly. They are all important and will have been put there for a reason.

Finally, remember to protect your hearing. Often it is more convenient to work on audio equipment using headphones rather than bulky loudspeakers to monitor the output. Do not play headphones too loudly and take them off when they are not strictly necessary. They can be distracting and certain fault conditions can result in sudden and very loud sounds. If you can hear a 'ringing' noise in your ears after using headphones, the sound has been too loud – loud enough to cause hearing damage. Reduce the volume and the time spent listening without a break when you use them next time.

Section Two: What to Collect

The earlier chapters of this book have shown that the complete B&O range is large and covers a wide range of different models. A collection based around such a complex and diverse range of objects will, by definition, be more difficult to organize and manage than one composed of a smaller set of simple items.

It is probably not possible to amass the whole model line. There are simply too many models, variations of models, colour options and accessories to make this something that can be realistically done, so it makes sense from the outset to decide on a 'theme' for your collection. This does not have to be cast in stone, of course, but it will help you in the early days to restrain yourself from taking a 'scattergun' approach and possibly making some poor buying choices that needlessly deplete your resources of money, space and time.

If you are planning to repair and restore the things in your collection, it is wise to stick to things you understand to start with. Some collectors collect only television sets, some only audio equipment of a certain era and some only portable models, for example.

Some collectors select items on appearance alone and use them as ornaments. Such collectors demand aesthetic perfection and completeness but are not greatly concerned by the prospects of anything in their collection ever functioning correctly again. A collection such as this can be very decorative but good examples of the most attractive well-known models can be expensive.

If you are starting a collection on a very small budget, you may just have to take what you are offered. This is a perfectly valid way of collecting and can yield some interesting results but it should be realized that it may take a while to assemble a complete and useful set of anything, or even to secure one item that truly works properly and is complete in every detail. Collectors following this route need to hone their skills of improvising, repairing and persistence, if they want to end up with anything other than tired cast-offs. This method of collecting, once mastered, can be very satisfying.

It is a good idea to remember that different products are used in different ways. A complex and expensive hi-fi system may have been used keenly for the first few months it spent with its original owner but once the novelty wore off, it may have just been used as a simple radio (or not used at all) for the bulk of the time. This means that for this class of equipment it is not that unusual to find relatively unworn examples of even some quite early models, and there is no need to settle for anything substandard; although, of course, those parts that deteriorate with time rather than use (drive belts, lubricants, adhesives, loudspeaker roll edges) will still require attention.

Television sets tend not to be used in this way. Larger models, in particular, may well have served two solid decades as a family entertainer, working several hours every night. This factor puts examples with a decent amount of life left in them at a premium, but one must not despair, as many of the sets mentioned in this book were exceptionally durable; it is surprising what is out there. Video recorders also suffer in a similar way, either acting as a babysitter to several generations of children or time-shifting the same soap opera by an hour or two, four times a week. One audio product that fits into this category is the BeoSound 9000. In their heyday it was not uncommon to see these used as a source of background music in the trendier sort of bar or hotel and so there are some very tired examples around. Check carefully.

What about individual models? There are some that collectors will always find desirable. The Beolab 5000 system, for example, the Beogram 4000 and the Beolit Colouradio range are the sort of things that every collector wishes they could have, and so good examples are always in demand and often expensive. Alternatives are not hard to find, however. The Beomaster 3000 or Beocenter 3500 make an excellent alternative to the Beolab; the Beogram 1200 series is in its way just as striking as the Beogram 4000 (and is much simpler too); and the earlier Beolit models (such as the original 600) are, in general, easier to come by than the later types and are still well made and well styled.

In the case of television sets 'the older the better' is a good general rule for the serious collector. Within a series, the smaller versions tend to be preferred because they are easier to accommodate; so, for example, in the first generation of colour models, the

A 1960s Beolit like this 600 is a good alternative to one of the later Colouradio models.

ABOVE: **The Beocenter 3500 is a convincing alternative to the more exotic separates systems of the period.**

BELOW: **The Beogram 1200 is a good choice if the 4000 is considered too complicated or too expensive.**

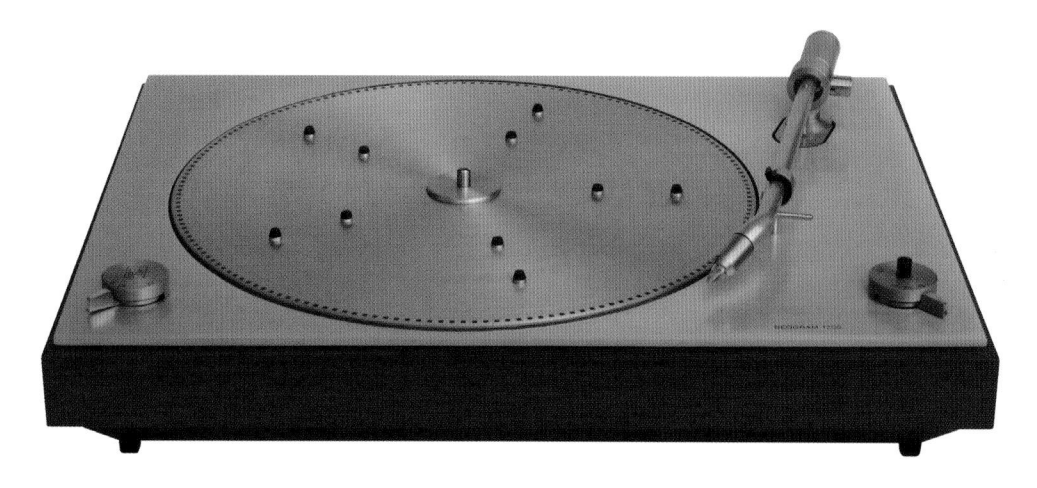

2600 is the favourite choice, although a tidy Beovision 3000 with a working chassis and a good tube is something few dedicated TV collectors would want to miss out on. As a television is a large item that tends to have a prominent position in the owner's home, the styling tends to be a little more restrained than is possible with audio products. For this reason many of the various models look very much the same. Exceptions to this rule will always interest collectors; the Beovision 3802 is a striking set, as is the AV 9000. A standard set in an unusual cabinet finish (e.g. the LX 2802 Whiteline) is also something to look out for. Similar comments apply to the Colouradio range: the more violent colours (purple, curry) are preferred by the collector today as there are fewer of them about.

Rarity is not always a good guide to use to judge if a particular model is worthy of a place in your collection. Rarity is often a sign that the model was not really wanted at the time, normally for good reason. Top B&O collectables, such as the Beomaster 900K, the Beomaster 1900 and the Beosystem 5000, all sold in huge quantities because they were worth having, a situation that remains true today. Age brings with it a different sort of rarity. As has been mentioned previously, some models sold in large numbers but have now all been 'used up' and so are very difficult to find in the sort of condition that makes owning them worthwhile. Others just slip into obscurity where, even though they still exist in reasonable numbers, they are just impossible to track down. This is particularly true of middle aged television sets (such as the Beovision 4402), which are seen by their owners as too old to have any value as a second-hand sale but too recent to be collectable. You will never see these in either dealers' windows or auctions of vintage equipment. Finding them is really just a 'random' process – you just have to be in the right place, at the right time.

Section Three: Where to Buy

If you would like to start collecting B&O equipment, first you have to find some. Listed below are some good sources where you can start looking.

Bang & Olufsen Official Dealers

Although B&O's own dealer network is geared to selling new equipment, some dealers offer a scheme called 'Second Life'. This offers second-hand models complete with a decent guarantee but, unfortunately, the choice is pretty limited, as only the most recent models are included. A better plan is to talk to the dealer and express an interest in the older ranges. B&O dealers are obliged to offer a high level of service and this may include part-exchange deals or the removal of old equipment. You might be able to buy some items on favourable terms, especially as it may otherwise cost the dealer money to dispose of them.

Independent B&O Dealers

There are now many firms and individuals who trade in used B&O equipment. Their stock tends to be made up from what the authorized dealers don't want, normally for reasons of age. Remember that when you buy from a dealer, you are paying dealer prices, so make sure you get a decent written guarantee (not just a scrap of paper with a mobile telephone number on it). A dealer who also runs a workshop encourages more confidence than one without.

Second-Hand Hi-Fi/TV Shops

Second-hand hi-fi shops tend not to stock much from B&O, as many of the models are seen as awkward to connect to other brands or as too complicated for a person dealing with many different makes to understand. This can mean that you get a good deal on something the shop owner is keen to move on, but check first; their lack of experience may mean that they overvalue some items.

Second-hand TV shops may have the occasional B&O set in their window that could be a good buy, if you can satisfy yourself that it has a reasonable amount of life left in it. Before deciding to buy, do make sure that it comes with a proper B&O remote control unit, not an aftermarket plastic alternative.

Internet Auctions

Be careful before bidding on any B&O products on an internet auction. Major cosmetic defects or missing parts may not be obvious in a small digital camera picture but be a real eyesore 'in the flesh'. Also be sceptical about a seller's description; 'good working order' could just mean that some of the lights came on when power is applied, so don't take such unqualified statements at face value. If you do end up buying something, always arrange to collect it in person from the seller's house and insist on a demonstration before parting with any money. Print out the seller's own description of the item before the auction ends and take it with you, so there can be no disagreement if the item does not match it when examined.

Internet auctions are valuable as a tool to check the approximate value of a particular model. Keep an eye on them, even if you never buy anything through them, and you will soon learn how much you will have to pay to secure what you would like.

Local Auctions

You may be lucky and find some B&O equipment amongst the house clearance items at a local auction. This can be a very economical way to buy, as there are unlikely to be many people there interested in similar things. Remember that there is no chance of testing anything or getting any idea of the history of the equipment, so assume the worst and restrict your bidding to a modest level. You may have to buy a whole large 'lot' to get the one thing you want but remember, you have to take it all away with you.

Audio Fairs

These are a recent innovation and can be an excellent way to expand your collection. Arrive reasonably early and have a good look around but avoid buying anything until nearer the end. Prices can tumble as the dealers realize that in a few hours what they haven't sold will have to be carried back out to the van. You won't normally get the chance to try anything out and are unlikely to get any sort of guarantee, so the prices should be low enough to offset this risk. Haggle!

Car Boot Sales

If you go often enough, you may be lucky and find B&O equipment at a car boot sale. Asking prices vary greatly, so have an idea in mind of how much the item you are looking at is worth, so you can gauge how realistic the seller is being. Ask lots of questions to determine if the equipment is something the seller has owned for a long time and used or is just a 'trade' item they are trying to make some money out of. Assume that what you are buying won't work and treat the seller's reassurances with a healthy level of suspicion, even if they offer you their telephone number for 'if should you have any difficulties'.

Trading Papers

These used to be an excellent place to look for used B&O bargains but their popularity has fallen now that internet auctions have become more widespread and easy to use. Some trading papers have a dubious reputation as an outlet for stolen goods, so tread carefully and keep you wits about you. Buy the paper regularly and see if certain telephone numbers crop up very frequently, as this will help you tell a regular dealer from a private individual making a single sale.

Local Newspapers

These are an excellent source of B&O equipment of all ages. Try to get the paper early and ring up immediately (keep trying if you don't get through) if you see something that you really want. A genuine seller in a local paper should set the equipment up in their home for you, so you can see that it works properly and, if this is not done, then find out why. Prices are normally very reasonable but model descriptions are often wrongly spelt either by the seller or the typesetter, so check this carefully in your initial telephone conversation.

Rubbish Dumps

Town councils are under pressure to recycle domestic waste and limit the amount of landfill they generate, so as a result it is sometimes possible to buy things from rubbish dumps, usually the 'interesting' looking things (which thankfully covers a lot of B&O's output) that the staff weed out and put on one side. Recent regulations in certain countries mean that electronic waste must be sorted separately, so you don't have to drag around the whole site looking in all sorts of odd places. Obviously assume that what you find will be defective in some way (although surprisingly this is not always the case) and ask first before taking anything. Look for cabinet damage, a lot of which is sadly done when the equipment is thrown on the pile, dragged along the ground or left out in the rain.

Friends/Colleagues

If you make it known that you are interested in B&O you never know what you may be offered. A B&O product is often a treasured possession and even when its original owner no longer needs it they may just put it away somewhere rather than discard it. Caring owners like this are frequently keen to see their old equipment go to a collector who will use it and care for it again. A note on a workplace notice board can work wonders and costs nothing.

Section Four: How to Buy

Buying complex electronic equipment second-hand can seem like a risky undertaking. Much has been written on the precautions to take when buying a used car or even a house, but there is little advice available for TV and hi-fi. In a way that can work to your advantage, as the electronics field tends to be a buyer's market, with the exception of a few well-recognized models for which there is always a strong demand.

The care you must take obviously depends on how much money you are going to spend. If the asking price is very low, then you must be prepared to 'take a chance' up to a point. It would be unreasonable to ask for an extended demonstration with lots of ancillary items for something being sold for less than the price of a round of drinks, for example. However, you should still keep your wits about you and your eyes open all the same. On the other hand, if the seller is asking a high price for what they have, it is clearly not acceptable to be asked to do the deal in a car park, where you have no idea of the seller's address, whether the equipment works or not and where you are allowed to only have a few minutes to inspect it. Use your own discretion and adjust your expectations accordingly in all cases; you will soon get the hang of it.

Obviously the ideal place to meet is in the seller's home or business premises. Clearly this is only possible in certain circumstances but 'high value' deals should be done this way. If the seller claims that the equipment is working, then you will want to see it demonstrated. Bypass any claims like 'I would show you it playing but I don't have a record' by bringing your own. With complete systems or models that offer multiple functions (e.g. music centres) be prepared to

Here are some examples of B&O serial number labels. This type was in used in the early 1960s.

This type was used in the late 1960s and early 1970s.

This design was used from the mid 1970s until the early 1980s.

Here is the latest style of label that you will find on the equipment covered in this book. On some models it may be larger and contain more information.

check everything. 'Good working order' means all of it. If, for example, the cassette deck is found not to work or the turntable stylus is broken, then 'Oh, I never use that bit anyway' is no excuse, the equipment is defective and should be valued accordingly.

Inspect the equipment for the serial numbers. These are normally printed on a small label that is stuck in place. Obviously, after many years the glue can dry out and the label falls off; this is unavoidable but recognizably distinct from labels that have been torn off (which leaves some paper and the glue behind) or defaced. Treat equipment like this with suspicion and ask lots of questions.

If something is being sold as 'working' then it is not unreasonable to expect a reasonably full set of cables and essential accessories to be included with it. If these are missing, then you are entitled to doubt how the seller knows that it is working; it is then down to the owner to explain the situation in more detail.

Some B&O equipment from the 2000 season and later uses a PIN code to act as a theft deterrent. This works in a similar way to the codes used to protect car radios and if you don't have it, then the equipment won't work. If you are buying something from this later period, ask a B&O dealer if it needs a code. If it does, be sure to get it and make sure that you are not given the 'service code', a special simple number that only allows for a few hours of use. For a charge, a B&O dealer can find out the PIN code for anything that needs it, but be aware that this information can be passed to the police and if what you have bought turns out to be stolen, you may at best lose it and at worst end up in court facing a charge of handling stolen goods. If the code is missing, ask the seller to apply for it; if they decline, then the safest thing is just to walk away.

Cheaper purchases can be approached in a less formal manner. In these cases what you do depends on what you want the item for. If you are buying as scrap to provide parts to mend something else, then all you have to do is to be reasonably sure that what you are looking at is a suitable donor and the part you want is still there and usable. Do this by asking questions, not by taking the thing apart while the seller waits, probably getting very annoyed! For things that you are hoping to use or restore, pay the closest attention to the cosmetic condition and completeness of the exterior and fittings; electronic/mechanical problems are far easier to resolve than gouges, dents, corrosion and missing minor controls. View a cheap purchase as something that you can afford to spend a reasonable sum on having repaired, not just something for which you have limited expectations.

If you do decide to buy what you have come to see, take time to prepare it carefully before taking it away. Assuming that you are using a car, bring some soft blankets or newspaper with you so that the cabinets are not marked on the way home. If your new purchase has a stand, bring the tools to take it off and dismantle it, if necessary, as this will make the whole task much easier and less risky. If original boxes are offered, certainly use them as they offer the very best of protection. Some models, such as record decks, CD players and cassette recorders, have transit locks that should be tightened down whenever the equipment is moved. Do not ignore them, as a lot of damage can be done to the suspension and cabinet otherwise. Record decks are the most fragile of all the main products and demand special care. Do not remove the pickup (this can damage it), instead wrap the arm gently in a piece of paper towel so that it cannot thrash around and break the stylus or mark the inside of the lid. Keep the machine flat at all times. With most models the platter is not secured, so remove it, as if it falls off, then it can damage the arm and spoil the cabinet finish.

Before you go, do remember to ask the former owner if they have any of the supporting paperwork that once accompanied your new acquisition. Owner's manuals are, of course, invaluable, as are original guarantee forms and dealer's receipts, as this gives you the exact age of the equipment, something that is always of interest to the collector. Another lucky find at this stage would be an original brochure, as these are expensive to buy and difficult to find on their own and will help you learn more about not only the model you have but also what else would work well with it. Some owner's manuals came complete with circuit diagrams, treasure these if they are still there, even if you can't read them. If you ever need to get

PRACTICAL COLLECTING   **187**

ABOVE: **Complete with its original paperwork, carrying bag, wall bracket and box, this Beolit 500 is quite a find.**

RIGHT: **Equally interesting is this Beomic BM 5 in its original box. If you are going to store items like this for the long term, take care that any plastic parts do not make contact with the expanded polystyrene tray, as a reaction can occur, which 'melts' the plastic. Put such parts in a plastic bag first.**

Beovision 8800.

the item repaired they will be of great help to the engineer and an easier repair is of course a cheaper repair.

Section Five: How to Inspect Equipment

Televisions

Televisions present an interesting case, for as well as checking that the set functions correctly, one must also assess how worn the picture tube is. This is not a matter simply of age or the amount of use the set has had; small differences in the manufacturing process are magnified over the years so that it is equally as possible to find the earliest colour models with tubes good enough to give really first-class results, as it is to find much later types that are completely useless.

With the set displaying a colour picture from a good aerial source, first turn the colour right down. The result should be a perfect monochrome picture with no overall colour cast. Sets with in-line tubes (the 20AX models and onwards) should display no colour fringes around any part of the picture and no odd coloured patches. Such things are excusable in the older models, as they can (and should) be adjusted out when the set is placed in the position where it is to be used.

Next, turn the colour back on and fully advance the contrast. The focus will inevitably deteriorate as you do this but the change should be small and the picture should remain of a reasonably high standard. If the picture becomes very blurred then the tube is reaching the end of its useful life.

There are other things to check too. Make sure all the set's basic functions work as they should (tuning, minor controls, etc.) and that the sound is clean and clear. Make sure the tuner works and that the set is not being operated from an external video source in an attempt to disguise a fault. There should be no hissing or regular ticking from inside. If there is, it suggests that the insulation around the high-voltage assemblies is defective and may give trouble in the

PRACTICAL COLLECTING 189

future. Another effect that betrays high-voltage trouble is a smell of ozone (a sharp, fresh smell) when the set is operating.

Check that the picture is big enough to fill the screen properly, especially in terms of width. If the set has a remote control, then also inspect this carefully and make sure that all the keys work (check the most frequently used ones, 1, 2, 3, 4, volume and standby) and that the battery cover is still there.

Video Recorders

These are very complicated and difficult to check completely. Bring your own tapes, so that the seller cannot fob you off when you ask to try the machine. The cassette should load smoothly and quickly. When playing back, make sure that the picture is stable and of an acceptable quality. A static horizontal line about a third of the way down the screen suggests worn video heads and, unless the machine is very cheap, it would be best to find another one. Try the winding modes; they should be brisk and run quickly up to the ends of the tape. Use the tape counter to assess how fast the machine is winding if you cannot see it. Sluggish winding modes indicate that a mechanical service will be needed, as does a loop of tape hanging out of the cassette when it is ejected.

Making a recording is a good test. When played back, the picture and sound quality should reach a high standard. Stereo recorders with hi-fi sound should select this automatically on playback. If the recorder selects mono sound, there is no sound or the sound is intermittent or 'spluttery', then the hi-fi sound heads are probably worn out. Replacement is not easy so the machine is probably best rejected.

Receivers/Tuners/Amplifiers

Because these are all completely electronic, they are quite easy to assess. Bring a pair of small loudspeakers or headphones with you (with an adaptor, if necessary), in case the seller claims not to have any.

Make sure both channels work (pay particular attention in case a crafty seller has plugged two loud-

RIGHT: **Beocord VHS 66**.

BELOW: **Beomaster 4400**.

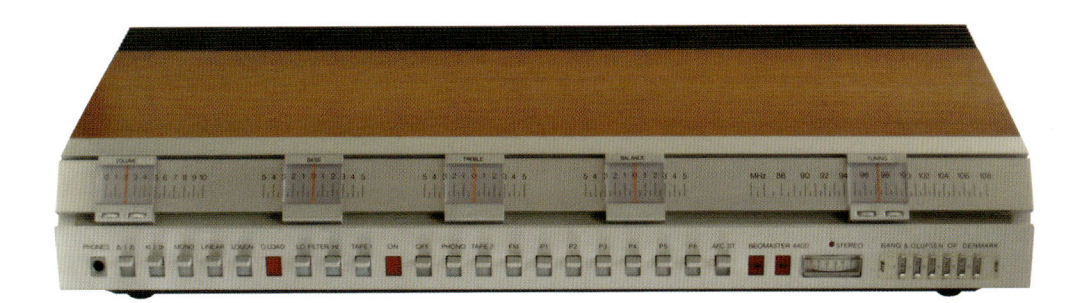

Beocord 5000.

speakers into one channel to disguise a defective amplifier) and that there are no humming or crackling noises in the background. Operate all the controls to check that they work smoothly and without introducing noise. Some models have loudspeaker switches or relays to switch in and out of several different pairs of loudspeakers. These are frequently troublesome, so make sure both loudspeakers work, whichever sockets they are connected to. After about five minutes play, feel the amplifier heat sinks. If they are very hot (perhaps on one side only), then the power amplifier is in need of some attention, budget for this in your offer.

An FM tuner should pull in a good few stations, even with only a metre or so of wire pushed into the aerial socket. In the case of mechanically tuned models, make sure that the tuning cord is not broken, as replacement is not easy. When inspecting types with digital tuners, make sure that the frequency display is accurate and that stations can be stored and recalled properly, as faults in the microcomputer control systems that support these functions can be very difficult to resolve.

Few problems are encountered with AM tuners. Some models mute the AM section if there is no aerial plug inserted; bear this in mind if the LW and MW bands are suspiciously silent. If you are keen on AM listening, consider tracking down the AM10 active antenna, as this is easy to set up and gives very good results.

Cassette Decks

A cassette deck in good working order should make recordings that are subjectively the same as the original programme. Dull, slurred, distorted or hissy results are all signs of mechanical wear or electrical misadjustment, they are not necessarily features of the cassette system. As with video recorders, the winding modes should be brisk if the mechanicals are in good condition. Budget for an overhaul, if they are not. Tape damage, especially near the start of each side, is another sign of advanced mechanical wear. Auto reverse recorders should be checked to ensure that they give similar sound quality (in terms of clarity, stability and tempo) in both playback directions.

Bring your own tapes to check the machine but don't bring anything with precious or irreplaceable recordings, just in case a fault in the recorder damages them. Bring a quality pre-recorded cassette to check the playback and a blank one for recording. Not all blank tapes are the same and so give different results. A standard type such as TDK AD or BASF LH-E1 should work well in any recorder, however. A second-hand classical recording produced by a respectable label

Beogram CD X.

(Deutsche Grammophon, Philips, etc.) makes an excellent test tape for playback and can be bought very cheaply in a charity shop.

Compact Disc Players

A good compact disc player will quickly read any disc that is inserted and swiftly be able to start playing back. A machine that needs several attempts or takes a long time is either out of adjustment or worn out. CD laser pickups are difficult to find and to fit, so it is wise to avoid any machine that has trouble reading the disc. Remember that with the Beogram CD 50 the disc goes in label side down and that these models are a little slower in reading the disc than the other types. A certain amount of erratic operation is par for the course with this model, although it should still play all the tracks on a disc and the sound quality should be excellent.

CD-R and CD-RW discs are not specifically recommended for older players, as they do not conform to any standard that was in place when they were made. That is not to say that older players will not play back such discs, it is just that for assessing a machine a fair judgement can only be made using 'proper' CDs.

Record Players

Record players tend to get scruffy if not looked after

Beogram 1700.

and the clear lids that most models in the B&O range have are vulnerable to scratches and marks, so look carefully. In most cases a hinged lid should stay open by itself when lifted but you may encounter worn or broken springs, which will stop this from happening.

The most delicate part of any record player is the pickup and stylus. B&O pickups and styli are very expensive and, therefore, any record player with these parts missing or broken is of limited value only. Earlier round S-type pickups have replaceable styli that can be bought fairly easily; the later square 'MMC' types

ABOVE: The SP-type pickups can be dismantled, allowing the stylus to be replaced.

RIGHT: Some Beograms, like this 1000 model, have a strobe ring built in. The bars should appear stationary at 33RPM when viewed under mains-powered lighting.

must be replaced as a complete unit if worn or damaged. B&O do not make pickups for record players any more but they have arranged for a company in America to produce compatible items. These are available through B&O dealers but they are costly and slightly different in appearance to the originals.

Assuming that the record player has a good pickup and can play a record, check that the turntable speed is correct and that the pitch is steady. You can use a strobe disc to do this or simply choose a record that you are familiar with.

Old Beogram models can sometimes take an age to get up to speed. If this is the case, the motor needs to be overhauled, which is not especially easy. Automatic models should drop the pickup precisely in the lead-in groove of the record; wear or misadjustment can mean that the arm either misses the record or starts well into the first track. Early tangential tracking models (Beogram 4000 series, 8000 series) are very complex and should be checked thoroughly. These types are sufficiently sought after that repairs are usually viable but all the same do not pay over the odds for examples with faults. Check the tangential drive system of the Beogram 8000, 6006 and 8002 very carefully. The motor should run silently and give perfect speed stability. Treat odd noises and speed 'jerks' with suspicion, as they can be difficult and expensive to resolve.

PRACTICAL COLLECTING   **193**

Beocenter 1800.

Music Centres/Combination Units

A music centre is simply a combination of any of the above units and so each part can be checked as if it were a separate item. Some music centres have faults in more than one section and so repairs can end up being very costly.

Loudspeakers

Of all the types of equipment so far discussed, it is loudspeakers that are most likely to be defective. Mention has been made in previous chapters of the foam roll edges that are used in many Beovox loudspeakers and you won't have to look at many before you encounter a truly rotten pair. To examine the roll edges you will need to take the grilles off, although this is not always practical. Early Beovox models like the Beovox 1600 (i.e. those where there is a large gap between the wooden cabinet and the edges of the cloth grille), were not made with foam edges, so in practice one does not need to worry too much. If the grilles are correctly fitted and show no obvious holes or dents, it is safe to assume that the cones behind are undamaged. The next generation of Beovox loudspeakers had cloth grilles tightly framed by bright metal trim. Some of these, for example the Beovox 2702, did use foam edges and need to be inspected. The recommended way to remove the grille is to

Beovox CX 50.

The simple and sturdy construction of the crossover networks used in Beovox loudspeakers makes repairs not too daunting, even for the inexperienced. The two grey components on the right are the potentially troublesome capacitors.

insert a palette knife covered by a thin rag between it and the metal trim, being careful not to mark anything. Uniphase loudspeakers all had simple clip-on grilles that can be easily removed, so there is no problem here. Not all had foam edges; the S35, for example, used rubber, whereas the S35.2 used foam, needless to say the former are the ones to go for. The grilles of the Beovox C and CX range just clip on and off. Inspect these carefully, even if the owner claims that they are not very old. Models newer than this are more difficult to inspect, so generally you have to take a chance. RL grilles, for example, are not removable. Fortunately, rotten foam is not an RL problem for anything other than the bass radiator of the RL 45 and RL 60. The grilles of Pentas are very difficult to remove and replace, so the best advice is to get someone who knows what they are doing to check them if you are spending a lot of money.

As well as the problem of foam rot, be careful in checking that all the drive units work properly. For this you will need to be able to listen to the loudspeakers, so make sure the seller has some appropriate equipment. You may have to bring your own, along with cables if you think these may be missing. Most Beolab active loudspeakers have an RCA line input socket, which will work from the headphone socket of a portable CD player if you have the right lead. This is a good way of testing them if a suitable B&O system is not available.

When the loudspeakers are playing, check that you can hear sound from each driver. Tweeters are easily damaged by amplifiers that are either defective or under-powered and the usual result is no output at all. Beolab 6000s and 8000s are well-known for having defective tweeters but for practical purposes treat all the drive units as suspect and take time to assess them all. Remember that if you find one defective drive unit, you will probably have to replace it along with its partner in the other loudspeaker to maintain the balance; this doubles the cost of the repair, so tread carefully.

Conventional Beovox loudspeakers in which one or more drive units seem to be underperforming may have a defective crossover network. Lack of treble or a dull, leaden sound is the usual result of this type of fault and the capacitors are normally to blame. Crossovers are simply constructed from large components, so replacing the faulty parts need not be that difficult.

Like all B&O products, loudspeakers should have their original serial number labels still in place. They are usually fitted on the back of the cabinet but may be placed underneath, if the model in question is designed to be placed in a manner where all sides of the cabinet may be visible. Most of the smaller models were made as a pair and, therefore, one would expect the serial numbers to be the same but this is not the case with some of the larger ones, such as the Beolab Penta, which were recorded as individual products. In all cases the type numbers should be the same, however.

It is always worth asking the owner if they have the original stands for the loudspeakers. They may have been put away and forgotten when the room was decorated or the furniture moved about so it is worth while jogging the memory, as loudspeakers with stands are much easier to place than those without.

PRACTICAL COLLECTING   195

Check the plugs and leads carefully before connecting any newly acquired loudspeaker to your own equipment. Badly fitted plugs or cuts into the wire can ruin an amplifier if they cause a short circuit, so make sure everything is sound. A previous owner may have removed the original DIN plugs and fitted different ones back on at a later date. If this has been done, ensure the phasing is correct, something that is most easily done by checking that the larger of the two plug pins is connected to the negative side of the woofer. A small battery connected across the plug, with its positive terminal to the smaller of the two pins, should make the woofer cone move towards the listener.

Section Six: Cabinet Care

Any collector naturally wants their collection to look its best and the first step in achieving this is to remove all the accumulated dirt that invariably settles on everything. B&O equipment, on the whole, has a durable finish and so can often be revived with a little care, although it is, of course, wise to closely inspect anything that looks like a deep scratch, gouge, rip or patch of corrosion at the purchasing stage, as not everything will clean off or polish out.

Always make sure that whatever you are cleaning is completely disconnected from the power source before you start work. The controls may be activated by accident or water may get spilt inside and result in a hazardous situation.

Start outdoors with a good dust over, using either a big soft brush or a soft cloth. Recover any small pieces that fall off at this stage. A compressed air supply or a vacuum cleaner with a 'blow' function is a very useful tool for removing dust, especially with television sets that attract a lot of it. Make sure the air flow is not so strong that it detaches trim, damages small parts (e.g. turntable pickups, video heads, loudspeaker cones), tears cloths or forces dirt into inaccessible places however.

Wood

Many B&O products of the collectable era have either wooden cabinets or wooden trim parts, which it is part of their appeal. From 1964 onwards all B&O's wooden finishes are unvarnished and, when presented correctly, should have a soft satin sheen.

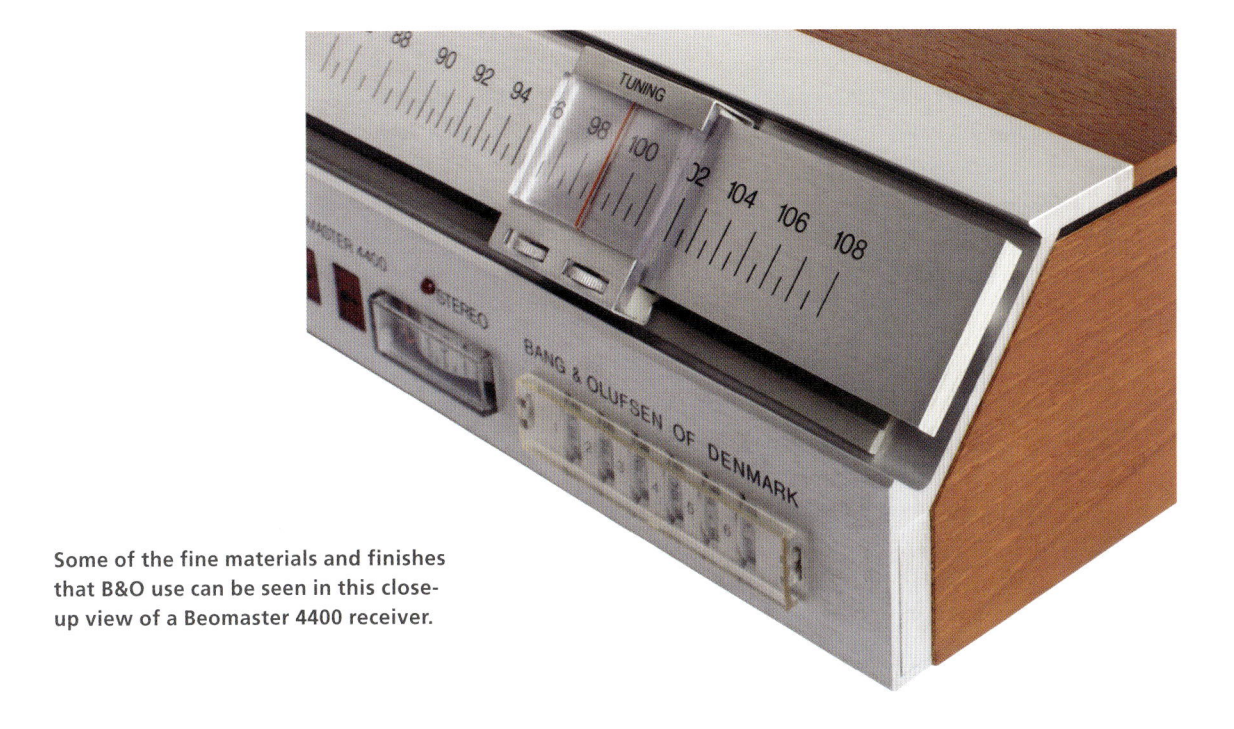

Some of the fine materials and finishes that B&O use can be seen in this close-up view of a Beomaster 4400 receiver.

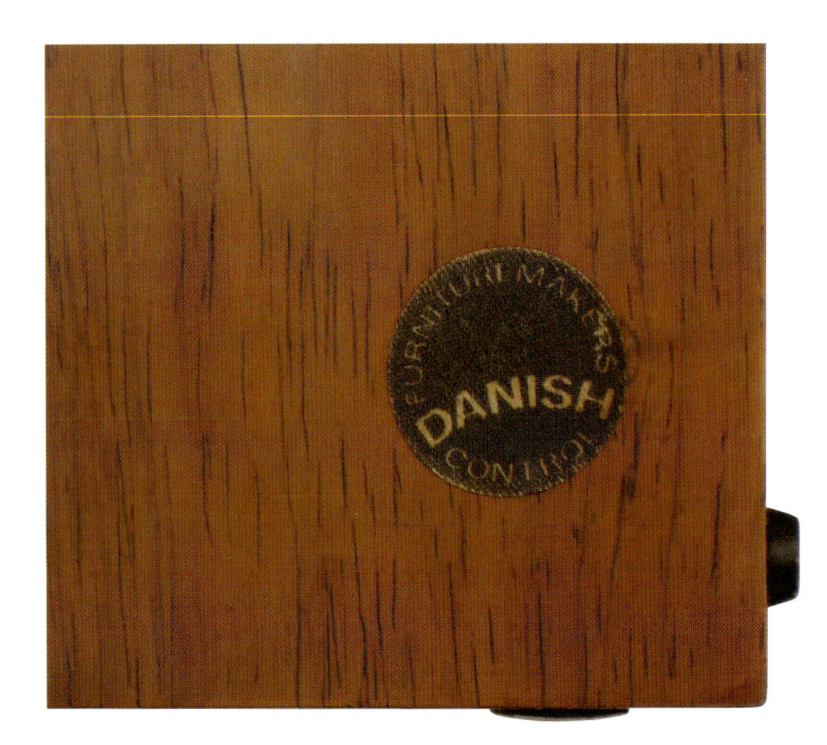

Some B&O models, typically those from the 1960s, have this Danish furniture makers' seal of quality.

A well-cared for example will only need a light wipe-over with teak oil on a rag to restore the lustre, repeated once a year to keep up the appearance. Sadly, this level of treatment is often not enough as the wood may have got wet, faded in the sun, been varnished at some stage or is just very dirty. In such desperate cases all is not lost. Firstly, remove as many of the working parts from the cabinet that you can. In some cases it may be easier to remove the wooden trim from the equipment rather than dismantle the whole thing. In any case, the idea is to keep the inevitable dust and dirt out of the works. If neither of these courses of action is possible, try to mask up any delicate finishes or openings. For loudspeakers, remove the grilles and take out the drive units one by one, noting which wires go where. This is particularly important, as the small particles from the steel wool that you will have to use will be attracted by the powerful magnets and jam the movement of the cones.

Start with a good wipe over with white spirit. Don't wet the finish too much, as it may dissolve the glue beneath, and don't rub out the natural colour of the wood. More often than not, a nasty brown gunk of dust, polish, varnish and finger grease will be lifted out as you carefully rub. Turn the rag and use clean spirit frequently. Once you've removed as much as you can, let the woodwork dry and examine it again. If the wood is smooth, all is well. If not, it can be flattened off with fine sandpaper over a block. Use this carefully and don't rub through the edges or round off the corners of the cabinet. After this, wipe the wood over with white spirit again and have another look. Neither of these methods will effectively remove dirt that has worked into the grain, so next, the wood can be rubbed down along the grain with fine steel wool. Add some more white spirit if there is a lot of dirt in the grain but do the last rub down with everything completely dry. The steel wool will quickly clog with debris, so turn it frequently. This stage will generate a lot of fine dust, so work outdoors and wear a mask if you are sensitive to that sort of thing. Clean away all the dust and steel wool particles once you have finished.

At this stage the wood will look grey and dull. A further wipe over with white spirit will give you an idea of what the colour will be like when the teak oil is applied, so do this lightly and assess what you have. If the colour is too light, it has probably faded, so use a wood stain carefully rubbed in to get it back to the right colour. Once this has completely dried (or if it turns out to be unnecessary), it is now time to apply the teak oil. This should be soaked into a cloth and carefully rubbed into the grain. Let it soak in, don't buff it. One or two coats may be required to get the desired finish but don't overdo it or the equipment will be oily to the touch and make your fingers sticky every time you use it.

Later models with plastic or metal cabinets often have small pieces of wood stuck on to give a pleasant appearance. The adhesives used were not very good and so the trim is often encountered peeling, split or missing altogether. The best thing to do, if this has

The cabinet of this Beovox loudspeaker has been neglected and is in poor condition.

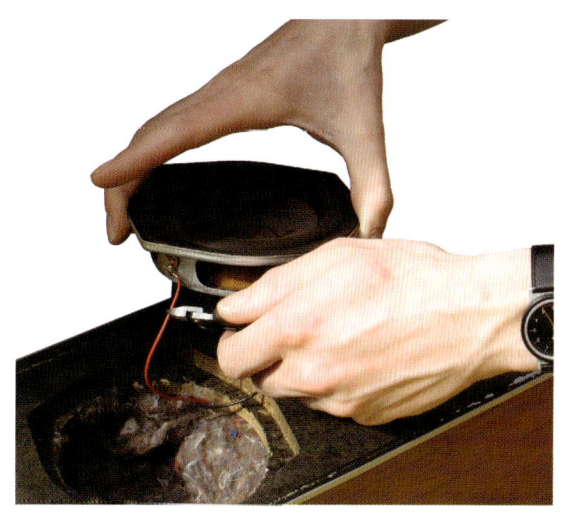

Remove as many of the working parts as you can before starting on any cabinet work.

Wiping the wood over with white spirit will remove a lot of grease and grime.

Smoothing the wood with abrasive paper wrapped over a block...

TOP LEFT: ...will leave a finish like this. The wood should be clean and smooth at this stage.

TOP RIGHT: A final rub-over with steel wool restores the original smoothness of the finish.

MIDDLE LEFT: Use wood dye if the colour of the wood has faded. Follow the makers' instructions to get the correct shade.

MIDDLE RIGHT: Once the dye has fully soaked in, teak oil can be applied to give a soft sheen and to protect the finish.

BOTTOM LEFT: After a lot of work, the loudspeaker cabinet looks presentable again.

happened to any great extent, is to strip off all the wood and replace it with new pieces. A cabinetmaker may be able to sell you off-cuts of veneer that are big enough to make all the sections you need and, in nearly all cases, only simple shapes with straight edges are needed. The old glue is tricky to remove from the cabinet parts but WD40 spray lubricant does dissolve some of the types that were used.

Clear Plastics

Anyone who has a turntable in their collection will straight away have a large area of clear plastic to look after. Similar materials are also found on tape recorders, receivers and some television sets. If you are lucky it will still be bright, clear and scratch free. If not, it will be a horrible hazy mess that makes the whole thing look scruffy and unattractive.

Start by wiping over the material with a clean cloth soaked in a weak solution of washing-up liquid and warm water. Obviously make sure there is no grit in the cloth and clean or change it if it picks any up during the cleaning operation. This may help a bit, although normally more work will be needed. A quality car paint preservative is ideal for restoring the finish of clear plastic; 'Mer' has proved to be especially good, if you can get it. Use it as you would to polish a car (follow the instructions) and you should find that some sheen returns and that the small scratches end up less visible as they are filled by the polish. Only do this where it is necessary; the inside of a turntable lid, for example, sees very little wear and normally needs only wiping over to remove the dirt.

TOP: **The lid of this Beogram 2000 is spoiled with marks and scratches.**

MIDDLE: **Apply the polish using a soft damp cloth in a firm, circular motion.**

BOTTOM: **After buffing, the finished result looks much more presentable.**

ABOVE: Clean parts, like this tuning scale, carefully – too much water or rubbing can make the printing faint or remove it completely.

BELOW: Be careful when cleaning textured metal finishes like the base of this Beolab 6000 loudspeaker.

Some models (e.g. the Beomaster 1900 series and the Beocord 8800V) have a special anti-scratch lacquer applied to the clear plastic parts. This was fine when new but with age it can start to lift and peel. This is unsightly and the only cure is to remove the lot. The best product to use again is Mer car polish. With a good deal of work, this will allow you to polish off the remains of the coating (it tends to come off in messy sheets and fragments) and leaves some protection for the plastic beneath.

The plastic contrast screen fitted to some MX TV sets (20 and 21in models only) should not be polished, it should simply be washed and carefully dried. It can be fitted inside-out to hide any unsightly marks and scratches. An MX set with a plastic contrast screen can be identified by a small notch in the middle of the screen at the bottom. Prise this out with your fingers and the screen will slip out of the channels at the sides. Bend the plastic a little to re-insert it.

Metal and Plastic

B&O metal finishes are well-protected and so can normally be wiped over with warm water and washing-up liquid to remove the dirt. Ingrained dirt that proves difficult to remove can be dissolved with an orange-based kitchen degreaser, although don't apply it neat; put a little on a damp rag instead. Don't rub too hard over printing or lettering, as some of it comes off quite easily, especially after many years of use. Use a toothbrush to get into the small corners but be careful not to get any water or residue inside the equipment.

Stickers and labels added by a previous owner can be an eyesore but take a measured approach to removing them. Most irritating of all are the station marker stickers that the BBC issued in the early 1970s. These take the form of little diamond shapes in either red or blue along with a frequency guide that is about the size of a postage stamp. These are very difficult to remove and carefully chipping them off with a fingernail is as much as you can do, clearing up the small traces of adhesive that are left behind afterwards with WD40 spray lubricant. Other plastic stickers can be softened with gentle heat from a hair drier, which

Treat finishes like this with the same care that you would with an expensive new car.

stops them tearing and tends to make them take their adhesive with them. Paper labels can be softened with water but make sure that this won't ruin the finish underneath.

Painted Finishes

Painted finishes can be split into two categories: glossy and textured. Glossy finishes (like that of the BeoVision Avant) are best wiped over and then polished with a quality car polish such as Mer. Textured finishes should not be polished, as this will result in shiny patches and a messy appearance. Instead, clean them with a stiff brush and warm water with detergent, then pat them dry.

Glass

Glass can be cleaned with products intended for cleaning windows. Be sure that the part in question really is glass, as some of these products can discolour plastics. The most obvious glass part to clean is a TV screen. Again, window cleaner can be used but don't choose a type that leaves an anti-static residue behind. If this gets inside it can cause problems with the high-voltage supply. Fine steel wool can be used to shift stubborn marks from the tube face but there is little that can be done about deep scratches or gouges, other than cleaning the dirt out of them so that they are less obvious.

Beovision LX and MX models with contrast screens present a different problem. Clean the contrast screen carefully, as it is often treated with an anti-reflective coating. This deals with the outside, but the inside surface and the tube face itself is often very dirty too. To clean this requires the contrast screen to be removed. This can be done at home but care must be taken. For the LX range, pull off the loudspeaker covers, then carefully prise out the four small square plastic covers that conceal the contrast screen mounting screws. Remove the screws carefully whilst holding the screen in place, then carefully remove the screen. The rubber dust seals may also fall out at this stage; re-fit these carefully in the cabinet, as they are important.

TOP LEFT: **Cleaning the Beovision LX contrast screen. First, remove the loudspeaker cloths...**

BOTTOM LEFT: **...then pry out the plastic covers over the screws.**

BOTTOM RIGHT: **Finally, remove the screws and carefully lift the glass contrast screen away.**

The MX range is constructed differently: the smaller sets with plastic screens have already been dealt with but the larger ones can be tricky. The mounting screws for the screen are hidden behind the shiny plastic strips above and below the screen. Start at the edges and gently remove them, being careful not to tear the loudspeaker cloth with the lower one. BeoVision Avant models and 21in MX sets with glass contrast screens need the cabinet to be partially dismantled to remove and clean the screen safely. This is best left to an expert if you are a novice.

Treat any contrast screen where the metal mounting strips are coming away from the glass with extreme care, as the glass may fall off the set and cause an injury. Various metal to glass bonding products are available but a replacement screen is a better option. If you do decide to attempt to re-bond the screen, check very carefully that the repair is sound before putting the set back into regular use.

Cloth

Loudspeaker cloths are best cleaned dry. Use a vacuum cleaner and a brush to remove the dust. Dirty patches can be brushed out with a clean toothbrush. Resist the temptation to wash the grille assemblies, even if they are easily removable, as warm water can soften the glue to the extent that cloths will become detached from the frame. If the cloths are badly stained, torn or holed it may be better to replace them. Choose a material that is as similar to the original as possible and remember that the sound has to get through, so hold the new cloth up to the light and make sure it is no denser than the original. Use contact adhesive sparingly to hold the cloth in place, once it has been cut to size.

Inside

If you are going to work on the equipment, it is good to start with the insides clean. Television sets, in particular, get very dusty and this will get everywhere when the set is dismantled. Working outside using an air blower is the best way to approach large-scale dust removal. Sometimes, after the billowing clouds have dispersed, a new-looking chassis will have appeared, which is very satisfying. Be careful, however, in case there are any loose parts that may be blown away and lost. For detailed work, a small brush is the best thing to use. The inside of any piece of equipment is easily spoiled by tobacco residue. There is not much that can be done about this but, if you are keen to get an 'as new' look inside and out, then the orange-based kitchen degreasing spray is the cleaner of choice. Use it sparingly on a damp rag, do only a very small area at a time and make sure everything is completely dry before applying the power to the equipment again.

Be careful cleaning loudspeaker cloths as many types, such as this coarse weave used in late 1960s Beovox models, were made specially and cannot be replaced easily.

Re-Attaching Loose Parts

During any cleaning operation, it is quite possible for some small parts to become detached. This is most often caused by old glue that has dried up and decomposed. Parts such as clear pointers on knobs, covers over VU meters and cassette viewing windows often come away cleanly and, provided that they are not lost, they are easy to re-fit. For parts that simply lay on top of each other, a thin double-sided tape is ideal; cut it neatly to size and the repair will be invisible. For parts that fit together in a way that the tape cannot be used (e.g. those that push together), use a tiny quantity of 'super glue', remembering that any surplus will squeeze out and potentially become visible when the joint is closed. This glue has a limited shelf-life once opened (about three months), so buy a fresh tube and only use quality branded types, if you want the repair to last.

Lifting wood trim and loose cloth is best re-fixed using contact adhesive. Follow the maker's instructions closely and apply sparingly.

Any equipment that has received amateur attention in the past will probably have screws missing or an assortment of the wrong types randomly used. This gives a bad impression and can make the equipment dangerous to use, if it falls apart unexpectedly. You can buy individual screws from an engineering supply shop but it is better to buy a kit of mixed lengths and sizes, then you will have what you need to hand and will also know what to re-order. Slotted 'cheese head' screws are the norm for early models with a mixture of 'pozidrive' and 'torx' types being used on the newer designs. All B&O products use metric threads throughout with M3, M4 and M5 being the most useful sizes.

Section Seven: Hints and Tips

Although B&O equipment is normally quite straightforward to use, it is possible to get far more out if you can gain some understanding of how it works and how to set it up properly. As has been previously stated, this is not a technical book but offered here is some simple advice that will set the beginner off to a good start.

B&O may still be able to supply instruction manuals for the equipment you have, so it is worth asking one of their dealers if they can sell you one. Alternatively, the B&O corporate website has a facility that allows you to download free copies of some of the later user manuals, once you have registered; the site address is: www.bang-olufsen.com.

How to Adjust Turntable Tracking Force

The amount of weight that bears down on the stylus of a record player as it tracks the groove of a record is a critical factor in determining the resulting sound quality of the system overall. Contrary to what may be expected, too little pressure causes more damage to both the record and the stylus than too much, so it is important to get it right.

B&O made a big point of how little weight their pickups needed to track correctly and the figures claimed are still commendable today. They should be taken as guidelines only, especially now that the pickups are old and the styli may be showing some wear. Even if you know the suggested setting for your particular pickup, it is best to arrive at a figure by testing and listening, as this ensures the most accurate results.

The first thing to do is to ensure that the adjustment scale on the arm is correct. In detail, the method varies but the principle remains the same for all types. With the pickup fitted (and any protective covers over the stylus removed or retracted), turn the adjustment dial or push the slider to the 'zero' position. Start the machine with a record on the turntable until the arm moves to the normal playing position. The arm should now float, showing no particular tendency to move up or down. If it does not, adjust the position of the balance weight until the arm balances correctly. With some designs this is easy, as there is either a rotating section at the rear of the weight to screw it in or out, or a small screwdriver hole at the back of the arm (usually accessible through a hole in the lid assembly) to adjust it. For those models that have a numbered ring at the front of the weight, simply turn the weight until the arm balances and then, holding the weight still, turn the ring until the zero is uppermost. If you can't see a way to adjust the balance point don't worry too much, you can work around it.

A good place to start determining what the optimum setting should be is about 1.5 on the scale. Set the control to this figure and play a record. Choose a good-quality pressing with lots of loud sounds and pick the loudest section. As the record plays, listen for harshness and distortion. If this is present, advance the setting slightly and try again, keep doing this until no more improvement can be heard, although be careful that you don't apply so much weight that the bottom of the pickup scrapes on the surface of the record. If no distortion or harshness is noted, you can retard the setting in small steps until the sound quality deteriorates, then advance it again until the best results are obtained. The best performance is normally obtained at the upper end of the range of values suggested by the manufacturer. Make sure the stylus stays clean throughout the procedure.

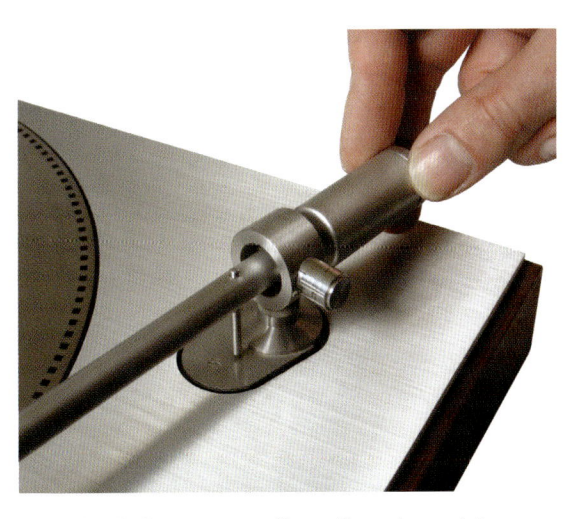

With the dial set to zero, first adjust the weight so that the arm balances. The method varies from model to model; a Beogram 1200 is shown.

With the arm balanced, advance the dial until the required tracking weight is achieved.

Cassette Tape Head Alignment

Because of the slow speed at which cassette tape runs, the precise angle that the head is set at in respect to the axis of the tape is critical. As the head wears, the adjustment can drift; so, if the machine performs poorly (dull treble, diffuse sound with poor stereo image), this is something to check. The adjustment screws are located right next to the playback head, so some dismantling is usually required to reach them; do this only if you know how to work safely.

Before you start, make sure that the heads are clean. Use meths or IPA (isopropyl alcohol) on a clean rag to wipe away any dirt and tape deposits. Work progressively at stubborn marks and don't be tempted to scrape at them. The pinch roller should also be clean and free of glaze or other contamination; if it is not, then medium-grade abrasive paper should be used to restore a clean non-slip finish. Most B&O cassette decks automatically de-magnetize their heads after every recording, so you don't have to worry too much about doing this. With the early models you can achieve a similar effect by setting the machine to record and then disconnecting the mains power.

To set the angle of the head you will need a reference. If you can get a proper alignment tape, then this is the best thing to use; however, these are expensive and not as widely available as they once were. Instead, substitute a good-quality pre-recorded cassette from a decent record label. The newer, the better is a good rule for choosing one, as standards were not adhered to very strictly in the early days of cassettes before the format became a serious hi-fi medium. Next, locate the adjusting screw. For normal decks the head will be mounted on a small strip of metal with a screw at each end, the one with a spring underneath is the one to adjust. Auto-reverse machines have a screw on each side of the head, one for each direction of play. Examine the mechanism closely before adjusting anything, as some designs also have adjustable guides in this area, which should not be moved under any circumstances. Once you have located the screw, fast-forward the tape to near the middle of one side, set the machine to play and listen to the results on stereo headphones. Turn the adjustment screw carefully until maximum treble is obtained on both channels. This may be easier to do if you turn the Dolby NR system off (if fitted). If the deck is an auto-reverse one, activate the 'turn' function and repeat the adjustment for the other direction of play. Once you are happy with the results, lock the adjustment screw(s) with a tiny drop of 'super glue'.

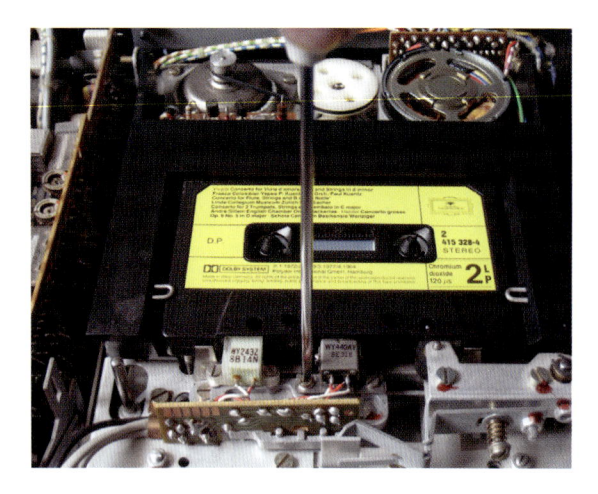

Aligning the tape head of a Beocenter 4000. Note the use of a quality pre-recorded cassette as a reference.

Do not attempt to adjust the heads of open-reel machines or video recorders using these simple methods, as they require skilled and specialist care. The original 1970s Beocord 5000 models are also very difficult to set up; special test tapes and jigs are needed to do it properly.

Positioning Loudspeakers

A loudspeaker is adjusted by altering its position in a room. Both the perceived bass and treble response can be changed dramatically, simply by altering the placement; so it is immediately obvious that both loudspeakers must be positioned in a similar manner. Treble is accentuated by moving the tweeters closer to the ear level of the listener and by directing them towards the listening position. In general, it is a good idea to maximize the treble effect using these methods. If this results in a harsh sound, there is probably a fault elsewhere in the system or in the material being played, which should be investigated.

There are no such simple rules in terms of bass. A loudspeaker will produce the most bass if it is placed in an alcove or a corner, and the least if it is stood in free-space. Designers tailor the response of each model to give the correct tonal balance in the situation where it is likely to be used; so, for example, small loudspeakers like the Beovox CX 50 or the S 22 work best when placed on sturdy bookshelves, whereas large models such as the Beovox RL 60 or S 80 should be fitted on their correct stands and placed on the floor, not too close to other furniture. Large loudspeakers like these can be tuned by moving them closer to, or further away from, a wall, as the closer they are the more bass will seem to be heard. Large active loudspeakers like the Beolab Penta have equalization switches that set how much 'bass expansion' is applied; the three settings are approximately for 'corner', 'wall' and 'free standing', in ascending order of bass output.

The distance apart at which the loudspeakers are placed is also important to ensure the best stereo imaging. A good basis from which to start is to imagine the two loudspeakers and the listener as the three corners of a triangle. Try to make all of the sides of the triangle an equal length but avoid pushing the loudspeakers too far into the corners of the room, if you can. Turn the loudspeakers inwards slightly, so that they point directly towards the listener. If you can,

Placing this Beovox RL 60.2 in a corner results in a sound that appears to have more bass.

PRACTICAL COLLECTING 207

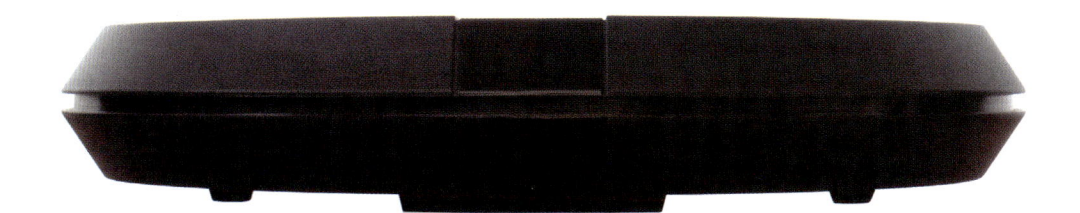

ABOVE: The Active AM Antenna 10 is an easy way to get the best AM performance from a Beomaster or Beocenter.

RIGHT: This type of indoor FM antenna has been a popular B&O accessory for many years.

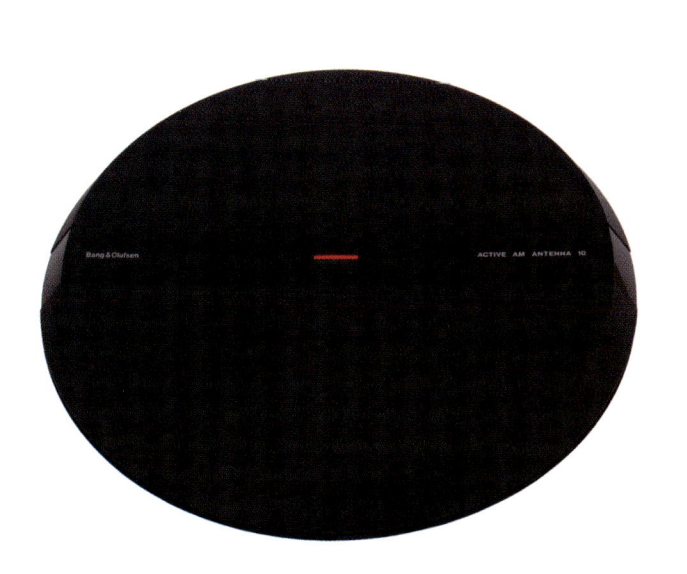

LEFT: Another view of the unusually styled but attractive Active AM Antenna 10.

BELOW: The FA1 rod antenna can also be used for AM reception.

locate the rest of the system close to the listening position and away from the loudspeakers, especially if you have a turntable.

Radio Aerials

Almost all Beomasters and Beocenters need an external aerial if the radio section is to be used effectively. This need not be as daunting as it sounds. The AM and FM tuners have separate input connectors, which should not join to a common aerial. B&O dealers sell a number of indoor aerials that can work quite well. For AM, either a simple length of wire placed as high in the room as is possible (about 2m is enough for local stations) or the active antenna AM10 can be used. You can, of course, use any length of insulated single-core cable as an AM antenna. Try pushing it into each of the two holes in the connector in turn to see which gives the best results. Keep all types of AM antenna away from TV sets, computers and such.

Another type of indoor antenna is the sort that uses flat wire in a T shape, which can be sufficient in strong-signal areas. If you have a choice, plug this type into the 300ohm FM aerial socket. If your set only has a 75ohm socket (this looks like a TV aerial socket), a radio shop should be able to sell you a matching transformer built into a special plug. Best of all, a folded dipole antenna mounted either on the roof or in the loft space can give first-class reception with virtually no background noise. These are usually matched to a 75ohm down lead, which should be plugged into the 75ohm FM aerial socket. If you want to run more than one set from the aerial, use a suitable amplifier to split the signal and don't just splice into the cable.

The AFC switch that you see on many B&O models is to aid accurate tuning on the FM band. Turning the AFC off makes the set very selective, so this should be done whilst tuning. Once a station is found, turn the AFC on and the receiver will lock firmly onto the signal. Sometimes the AFC button is labelled '-AFC'; this shows that you press it to turn the system off. Some models also have 'silent tuning'; the button for this is labelled 'ST' and its function is to mute the FM receiver, unless a strong station is tuned in. This effectively eliminates the noise that is normally heard whilst tuning between stations but, if the aerial signal is weak, it may block everything. Check this if there seems to be no reception at all.

Receivers like this Beomaster 2200 have a silent tuning (ST) function combined with the AFC switch. Check this if the FM band appears to be silent.

CHAPTER SEVEN

SYSTEM BUILDER

If you are very lucky, every time you buy some B&O equipment to add to your collection, it will come as a complete system with all the original components, the correct loudspeakers and maybe even the original special furniture to stand it all on. In practice this seldom happens, especially if your budget is limited. It is, however, not too difficult to put a system together if you buy all the units separately, even if it can sometimes take a while to get the full outfit. Remember that most models were available in a choice of finishes (typically teak, rosewood, oak or white for the early types and white as an option for the later models), so make sure you ask what colour a prospective purchase is if you want to match it to existing equipment and you are buying unseen or making a long trip to view.

Many models were originally sold as a part of a complete system. These can be easily recognized as each unit shares the same number; for example, the Beomaster 1900 was intended to be partnered by the Beogram 1900 and the Beocord 1900. In most cases, the loudspeakers did not share a common number, as a choice of sizes and styles were usually available but there are some exceptions: the Beovox 5000 with the original Beolab 5000 system, and the Beovox 1001 with the Beomaster 1001 for example.

Datalink Systems

From the mid-1980s onwards, many B&O models were fitted with seven-pin DIN Datalink connectors. This new method of interconnecting hi-fi components carried control data as well as sound and made setting up a sophisticated system very easy. Most models with these connectors can be interchanged freely, although a system comprised of matching units is usually preferable.

The Beomaster 1000 and Beogram 1000 were clearly made to work well together.

The following tables show the 'master' units in the left-hand column and the recommended sources on the right. Later models, such as the BeoSound 9000 and BeoSound Century also feature seven-pin connectors and most Datalink equipment can be used with these; however, the data connections are not always enabled so remote control of the extra sources may not be possible.

Some Datalink Beomasters and Beocenters are equipped with a socket marked AUX/TV. These can be connected to a Beovision TV set fitted with a suitable seven-pin socket (LX '02' series and newer, MX 3500/5500 and newer in general) and can then interact with them and other B&O equipment (e.g. a suitable Beocord video cassette recorder) that is connected to them. The Beocord VHS 91 and VHS 91.2 also have a Datalink connector and can be connected to the TP1 or TP2 socket of a Beomaster or Beocenter and then operated in the same way as a normal tape deck. The original LX TV sets and the Beocord VHS 66 could also be equipped with a special Datalink module.

BEOMASTER-BASED DATALINK SYSTEMS

Beomaster	Beogram	Beocord	Beogram CD
2000†	2000	2000	CD X†
3000†	3000	2000	CD X†
3300	3300	3300	3300
3500*	3500*	3500	3500
4500*	4500*	4500	4500
5000	5000, 5005	5000 (late)	CD 50
5500	5500	5500	CD 50, 5500
6000 (late)†	6000, 6006, 6002, 8002	6000, 6002, 8004, 9000	CD X†
6500*	6500*	6500	6500
7000*	7000*	7000	7000
8000†	8000, 8002	8000, 8002, 8004, 9000	CD X†

BEOCENTER-BASED DATALINK SYSTEMS

Beocenter	Beogram
8000	8000
8500	8500
9000	3000, 9000
9300*	7000*
9500	9500

Notes:

Models marked with * do not have a pre-amplifier for a turntable and so must be used with the turntables that are also marked *. Later models with seven-pin AUX sockets will also work with this type of turntable only.

Beomasters marked with † do not have a Datalink connection for a compact disc player. For these models, the Beogram CD X (also marked †), CDX 2 or a CD 50 fitted with the IR sensor kit are recommended.

The Beogram CD 50 could be fitted with an infrared receiver, the remote control unit that went with this (the CD Terminal) could also be programmed internally to operate some functions of the Beomaster 3000, 6000 and 8000, or of the Beocenter 7007 and 7700. All these optional accessories are rare, however.

Older Systems (Pre-Datalink)

Older models present a more difficult proposition. Frequently only a partial system was produced, so a complete matching system of all the available sources cannot be assembled. There was also a change in the signal level that a Beomaster sent to a Beocord for recording. This change came in the early 1980s and started with the Beosystem 8000. In general, only Datalink models feature the new higher level but problems may still occur if these are used with older equipment. An early (pre-Datalink) Beomaster or Beocenter may not produce sufficient output to drive a CD or MiniDisc recorder or computer sound card, either. Aside from these limitations, early equipment can be interchanged as freely as is possible with later types.

Some of the potential problems are highlighted in the following tables, which make suggestions of systems you may wish to try. The column on the left shows the component that houses the amplifier; the next column(s) show some possible choices for source equipment, which is of the correct period and will

BEOMASTER-BASED SYSTEMS

Beomaster	Beocord	Beogram
800	900, 1101	1202, 1203
900*	1500†	1000VF*
901	900, 1101	1100, 1200, 1202, 1203
1000	1500†	1000V, 1800 (early)
1001	1200†, 1700	1000V, 1001
1100	1100	1100, 1202, 1203
1200	1200†	1200, 1202, 1203
1400 (early)	1200†, 1800†	1000V, 1800 (early)
1500	1500 (late)	1102, 1500
1600 (early)	1200†, 1800†	1000V, 1200, 1800
1600 (late)	1600	1700, 2200
1700	1700 (late)	1700. 2200
1900	1100, 1500, 1900, 2400	1900, 1902, 2200
2000 (early)	1700 (early), 2200	1202, 1203, 2000 (early)
2200	1900, 2400, 5000 (early)	1902, 2200, 4002
2300	1500, 1900, 2400	1700, 2200, 2202
2400	1500, 1900, 2400	1902, 2200, 4002
2400-2	2400	2400, 2402, 4004
3000/3000-2	1200†, 1800†, 1700, 2200	1800, 3000 (early), 4000
3400	1700 (early), 2200	1900, 3400
4000	2200, 5000 (early)	3000 (early), 4000, 4002
4400	2200, 5000 (early)	1700, 4000, 4002, 4004
6000 4channel	5000 (early)	4002, 4004, 6000

BEOLAB-BASED SYSTEMS

Beolab	Beomaster	Beocord	Beogram
1700	1700 (early)	1700 (early), 2200	1202, 3000 (early)
5000	5000 (early)	1500†, 1800†, 2000 D/L†, 2400†	1800, 3000 (Thorens), 3000 (Acoustical), 4000

OPEN-REEL BEOCORD BASED SYSTEMS

Beocord	Beomaster (tuner)	Beogram
2000 D/L†	5000, Beolit 1000	1000V, 1800 (early)
2400†	5000	1800, 3000 (early)
1600† *	1700	1200 + GF4*

BEOGRAM-BASED SYSTEMS

Beogram	Beomaster (tuner)	Beocord
1500 (record player)	5000 (early), Beolit 500 (early)	900, 1200†, 1500†

BEOCENTER-BASED SYSTEMS

Beocenter	Beocord	Beogram
1400	Tape 2: 900, 1101	1202, 1100
1500	Tape 2: 900, 1101	1202, 1100
1600	Tape 2: 1100	1202, 1100
1800	1100	Integrated (1100)
2600	Tape 2: 1500, 1900	1500, 1902
2800	1500, 1900	Integrated (1902)
3300	1500, 1900, 2400	Integrated (1902)
3500	1200†, 1700 (early), 2200	Integrated (3000)
3600	Tape 2: 1100	Integrated (1100)
4000 (early)	Tape 2: 1900, 2400, 5000 (early)	1700, 1902, 4002
4000 (late)	(two integrated tape decks)	1800, 2000, 3000 (all late)
4600	Tape 2: 1500, 1900	Integrated (1902)

Notes:

Models marked * lack a pre-amplifier stage for a turntable pickup and Beograms marked * include a suitable circuit. Beolit portable radios with a record player input (e.g. the early 500, the later 600 and 700 and the 1000) also all need a record player with a built-in pre-amplifier.

† indicates tape recorders that are open-reel machines; those not marked use standard cassettes.

Beocenter models that include a turntable are listed with 'integrated' in the Beogram column along with the nearest separately available equivalent model. Those models that include a tape deck but can also be permanently connected to a second machine, have suggested choices listed as 'tape 2'.

work with it. Music centres are not listed if they are self-contained and not designed for the permanent connection of other sources. Such models include the Beocenter 7000 and 7002, where the insertion of a plug into the auxiliary socket blocks all the other sources. The Beocenter 7007 and 7700 have a TP2 key that selects four RCA type sockets to which any modern audio source or recorder can be connected.

appearance. The later universal Beograms were frequently supplied without a pickup to reduce the list price.

Although the Beogram CD 50 has Datalink and is primarily designed for operation in a B&O system, it can also be used with other non-Datalink equipment when fitted with an optional remote control kit. As mentioned previously, however, this is a rare item.

Universal Models

Some B&O models were designed without a particular system in mind. Instead they were intended for the selective updating of older installations or to improve an otherwise non-B&O system. They don't have the Datalink system and sometimes don't use DIN connectors, which are otherwise universal in the B&O range. The 1500 system of the late 1970s is a complete set of universal products. Even though they bear the same number, they do not really match at all in

Power Link and Speaker Link

Power Link is a special eight-pin connector for the easy connection of active loudspeakers. They can be interchanged freely as tastes and budgets allow. The Power Link system (in slightly modified form) was still in use at the time of writing and there are now many more models that use it. Any model made after the 2000 season that uses external loudspeakers needs Power Link types, ordinary 'passive' loudspeakers cannot be used.

An earlier system called Speaker Link, which uses

UNIVERSAL BEOGRAM MODELS

Beogram	Similar to	Connector
1500	1902	DIN
1800 (late)	2000 (late)	DIN
RX	2000 (late)	DIN
RX2	2000 (late)	RCA
TX	6002	DIN socket
TX2	5005	DIN

OTHER UNIVERSAL MODELS

Name	Similar to	Connector
Beogram CD X	–	RCA
Beogram CDX 2	Beogram CD 3300	RCA
Beogram CD 50 + IR kit	–	DIN socket/RCA
Beomaster 1500	Beocenter 2600, 2800	DIN sockets
Beocord 1500	Beocord 1900	DIN socket

POWER LINK/SPEAKER LINK AUDIO SOURCES

Audio Source	Power Link	Speaker Link
Beocenter 2300	●	
Beocenter 2500	●	
Beocenter 8500	●	●
Beocenter 9000		●
BeoCenter 9300	●	●
Beocenter 9500	●	●
Beomaster 3500	●	●
Beomaster 4500	●	●
Beomaster 5500		●
Beomaster 6500	●	●
Beomaster 7000	●	●
BeoSound 9000	●	
BeoSound Ouverture	●	

POWER LINK VIDEO SOURCES

Video Source	Power Link	Speaker Output	Surround Sound
AV 9000			●
BeoCenter AV 5	Only with surround sound		Optional
BeoVision 1	●		
BeoVision Avant	Only with surround sound		Optional
Beovision LX 4500	●	●	
Beovision LX 5000	●	●	
Beovision LX 5500	●	●	
Beovision LX 6000	●	●	
Beovision MX 3500	●	●	
Beovision MX 4000	●	●	
Beovision MX 4002	●	●	
Beovision MX 5500	●	●	
Beovision MX 6000	●	●	
Beovision MX 7000	●	●	

two additional pins on an otherwise conventional DIN loudspeaker connector, is also fitted to some models to control the display panels of those Beolab loudspeakers that have them.

Any audio sources manufactured before 1987 and not listed here can drive loudspeakers with Speaker Link, when connected via a shielded cable with standard two-pin DIN loudspeaker plugs.

There are no video sources with Speaker Link. Optional surround sound modules have Power Link connections only.

POWER LINK/SPEAKER LINK LOUDSPEAKERS

Loudspeaker	Power Link	Speaker Link	RCA Line In
BeoLab 1	●		●
Beolab 2500	●		
Beolab 3000	●	●	●
Beolab 4000	●		●
Beolab 4500	●	●	●
Beolab 5000	●	●	●
Beolab 6000	●		●
Beolab 8000	●		●
Beolab Penta 1		●	●
Beolab Penta 2	●	●	●
Beolab Penta 3	●	●	●

Loudspeakers

Choosing loudspeakers is very match a matter of personal taste. All the models sound slightly different and are frequently the most difficult items to integrate with the rest of the room décor. For the collector, the best loudspeaker choice is one that is both suitable and appropriate. A suitable loudspeaker is one that matches the power of the amplifier with which it is used. Loudspeakers with low power ratings can be over-run by using them at high-volume settings with an amplifier that is too powerful. Using an insensitive loudspeaker designed for large amplifiers with an underpowered system can also lead to problems because the 'clipping' (distortion that the amplifier produces as it reaches the limits of its abilities) can damage the loudspeaker's tweeters.

Many Beovox loudspeakers are marked with their power rating. Others have clues in the model name, Uniphase (S series) and the earlier RL models have a number included in the model name that is the approximate power handling in watts; for example, the Beovox S 55 is nominally a 55W loudspeaker, while the Beovox RL 140 can handle 140W.

To summarize, choose a loudspeaker that can be positioned well in the room where it is to be used and that matches the amplifier that you have.

Rather than listing loudspeaker/amplifier combinations, the following tables show the output power of some popular B&O models. The figures given are for

OTHER MODELS (ALL UNDER 10W)

Beogram (record player)	Beocord (open-reel)	Beolit
1500 (3 speed)	Stereomaster	500 (early)
1500 (2 speed)	2000, 2000 D/L	600 (Colouradio)
	2400	700 (Colouradio)
	1600	707
		1000

BEOMASTER MODELS

Model	<10W	10–20W	20–30W	30–40W	40–50W	50–60W	>60W
900K	●						
900M	●						
900RG	●						
901		●					
1000		●					
1001		●					
1100			●				
1200		●					
1400		●					
1500			●				
1600 (early)		●					
1600 (late)			●				
1700 (late)			●				
1900				●			
2000 (early)					●		
2000 (late)				●			
2200					●		
2300				●			
2400				●			
2400-2				●			
3000 (early)				●			
3000 (late)				●			
3000-2					●		
3300					●		
3400			●				
3500					●		
4000						●	
4400							75W
4500					●		
5000 (late)						●	
5500						●	
6000 4channel				●			
6000 (late)							75W
6500						●	
7000						●	
8000							150W

BEOCENTER MODELS

Model	<10W	10–20W	20–30W	30–40W	40–50W	50–60W	>60W
1400		●					
1500		●					
1600			●				
1800			●				
2000			●				
2002			●				
2100			●				
2200			●				
2600			●				
2800			●				
3300				●			
3500				●			
3600			●				
4000 (early)				●			
4000 (late)			●				
4600			●				
5000				●			
7000				●			
7002				●			
7007				●			
7700				●			
8000				●			
8500				●			
9000				●			
9300				●			
9500				●			

BEOLAB MODELS

Model	<10W	10–20W	20–30W	30–40W	40–50W	50–60W	>60W
150							150W
200							150W
1700		●					
5000						●*	120W†

* When used as a stereo amplifier.
† When used as a mono amplifier.

continuous RMS (1kHz sine wave) power into the rated impedance with both channels driven. In the late 1980s, the method of measurement that was used to give the headline figures in the catalogues changed to a rather dubious technique that resulted in a doubling of all the figures. This is best dismissed as marketing hype; the more realistic earlier method has been used here.

Televisions and Video Systems

Once video recorders became popular, B&O produced a sometimes quite large range to match their various TV sets. Early Beocord video cassette recorders could be used with any TV set, but for remote control they needed the Beovision Video Terminal handset that was supplied with some of the more expensive televisions. V and VX series recorders used the infrared receiver in the TV set and so could not easily be used on their own or with older models.

The first B&O video recorder, the open-reel Beocord 4000, is a complex machine that is designed to work with its matching camera and monitor, so it is difficult to use in any other configuration. In addition, it does not work like an ordinary video cassette machine, as it lacks a tuner, timer and sometimes an RF modulator, so it is not listed here.

All the TV sets and video recorders in the table below, with the exception of the L/LX models and the MX 3000, can be interchanged as they all use the same remote control system. If the television has an AV connector (SCART or DIN), then this should be used to make the signal connection to the video recorder, especially if the equipment is stereo.

Older Beovision TV sets may not be completely compatible with a video recorder. If there is a problem, it will show itself as distortion at the top of the picture. Modifications were issued to cure this effect in all models but they need specialist knowledge to implement. Many sets will have been converted already. If you want to use an older set without an AV connector with modern equipment that does not

VIDEO RECORDERS THAT USE THE BEOVISION VIDEO TERMINAL REMOTE CONTROL SYSTEM

Beocord VTR	Suggested Beovisions
8800V†	5500, 7700, 8800, 9000
8802V†	5502, 7702, 8802, 9002
VCR 60†	5102, 7102, 8102
VHS 63*	5002, 5100, 5102, 7100, 7102
VHS 66	5502, 7702, 8802
VHS 80	5502, 7702, 8802, 9000
VHS 82*	MX 2000, M 20
VHS 82.2*	MX 2000, MX 3000, M 20
VHS 90	7802, 8902, 9002
VHS 91*	L/LX 2500, L/LX 2800, 5502, 7802, 8902
VHS 91.2*	L/LX 2502, L/LX 2802

Notes:
Models marked * can be switched or adapted for use with direct control via the SCART connector, a system that became the norm once the VX models were introduced.

Models marked † use the Video 2000 (V2000) format and cannot play VHS tapes. All other models use the VHS format.

VX VIDEO RECORDERS

Beocord VX	Beovision
5000	L/LX 2502, L/LX 2802, MX 5000
5500	LX 4500, LX 5500, MX 3500, MX 5500
7000	LX 5000, LX 6000, MX 4000, MS 6000, MX 6000, MX 7000
7000 Beo4	MX 4000 Beo4, MX 6000 Beo4, MX 7000 Beo4

OTHER LATER VIDEO RECORDERS

Beocord	Beovision
V 3000	LS 4500, LS 5000, LS 5500, LS 6000, LE 6000
V 6000	LS 5000, LS 6000, LE 6000, ME 6000
V 8000	MS 6000, MX 6000 Beo4, MX 7000 Beo4, AV5, BeoVision 1

contain its own RF modulator (e.g. some DVD players and set-top boxes) you can buy an external RF modulator as a separate unit.

The VX series of video recorders are very sophisticated and need the correct TV set to work properly. The digital picture features of the Beocord VX 5000 are only generally available if an '02' series LX TV or an MX 5000 is used with them. Older versions of these sets (the L/LX '00' range and the MX 4500) could also work with the VX 5000 but full functionality was only available if a software update had been fitted to the TV.

Later TV models share their tuning and clock data with the VTR. This will work with the VX 5500 and VX 7000 VTR models and the VX 5000 if the software variant is 2.0 or higher. Later VX 7000s are optimized for use with the Beo 4 remote control; these models have 'Beo4' in the model number.

The above table gives suggestions for the correct equipment to use. Most of these models may be interchanged but some of the minor functions of the VTR may not be operable if the TV software does not match.

Remote Controls

Many models can be used with remote control, the following tables show the correct types to use. Beomasters 5500, 6500 and 7000 can be operated using a simple remote control but for timer programming the correct Master Control Panel (MCP) or a Beolink 5000 or 7000 (as appropriate) is needed. The Beolink 5000 and 7000 can also be programmed to operate any equipment that uses the Beolink 1000 but the two-way functions will not be available.

ULTRASONIC REMOTE CONTROL

Beomaster	Remote control
6000 4channel	Beomaster 6000 Commander
2400, 2400-2	Beomaster Control Module

Beovision	Remote control
Beovision 6000	Beovision 6000 Commander
Other models	Beovision Control Module

INFRA-RED REMOTE CONTROL

Beomaster	Remote control
3000	Terminal 3000
3300	Terminal 3300, A Terminal, AV Terminal, Beolink 1000
3500	Beolink 1000
4500	Beolink 1000, Beolink 5000
5000 (late)	Terminal 5000, MCP 5000
5500	A Terminal, AV Terminal, Beolink 1000, MCP 5500
6000 (late)	Beomaster Terminal
6500	Beolink 1000, Beolink 5000, Beolink 7000, MCP 6500
7000	Beolink 1000, Beolink 5000, Beolink 7000
8000	Beolab Terminal

Beocenter	Remote control
7000, 7002	Beocenter Control Module
7007, 7700	Beocenter Terminal, MCP 7700
8000, 8500	Beolink 1000
9000	A Terminal, AV Terminal, Beolink 1000
9300	Beolink 1000
9500	Beolink 1000, Beolink 5000, Beolink 7000

Beocenter/BeoSound	Remote control
2300, 2500	Beolink 1000, Beolink 5000, Beolink 7000
2300 (later model)	Beolink 1000, Beo 4
AV 9000	Beolink 1000, Beolink 5000, Beo 4 (later models)
BeoSound Ouverture	Beolink 1000, Beo 4 (later models)
BeoSound Century	Beolink 1000, Beo 4
BeoSound 9000	Beo 4

INFRA-RED REMOTE CONTROL contd.

Beogram CD	Remote control
CD 50	CD Terminal + IR kit (optional fitment)

Beovision	Remote control
33XX (manual tuning)	Beovision Video Terminal (black with yellow text)
77XX (digital tuning)	Beovision video Terminal (grey with orange text)
LE/LS/ME/MS series	Beolink 1000
L/LX 2500, 2800	V Terminal, AV Terminal, Beolink 1000
L/LX 2502, 2802	Beolink 1000
Other LX models	Beolink 1000, Beolink 5000
MX 1500	Beolink 1000
MX 2000, M 20	Beovision Video Terminal (grey with orange text)
MX 3000	Beolink 1000
MX Beo4 models	Beo 4
Other MX models	Beolink 1000, Beolink 5000
AV 9000	Beolink 1000, Beolink 5000, Beo 4 (later models)
Avant, AV5	Beo 4
BeoVision 1	Beo 1, Beo 4

Beocord Video	Remote control
8800V, 8802V, VCR 60	Beovision Video Terminal (black with yellow text)
VHS 80, VHS 90	Beovision Video Terminal (grey with orange text)
VHS 63, VHS 66	Beovision Video Terminal or via AV socket
VHS 91, VHS 91.2	Beovision Video Terminal or via AV socket
VHS 82, VHS 82.2	Internally configurable for all systems
VX models	Via AV socket or Beolink 1000 with VX sensor
V 3000, V 6000, V 8000	Via AV socket

INDEX

20AX colour tube 87, 88, 188
30AX colour tube 95, 96, 131
45AX colour tube 129, 153
A Terminal 138, 144
Acoustical 37
active loudspeakers 158
Aiwa 102, 134
AM 10 antenna 190, 207, 208
ambiophonic sound 53
Arena 22
Ascom 172
AV 9000 home cinema 161, 162, 163, 166, 168, 182
AV Terminal 138, 139
Bakelite 13
Bang, Jens 63
Bang, Peter 9, 10, 11
Beo 1 177, 178
Beo 4 167, 168, 177, 219
Beo- prefix, origin of 14
Beocenter 1400 74
Beocenter 1500 74, 78, 80
Beocenter 1600 74, 75, 78, 102
Beocenter 1800 75, 76, 193
Beocenter 2000 99, 100
Beocenter 2002 99, 100
Beocenter 2100 100
Beocenter 2200 100, 101
Beocenter 2300 163, 169, 170
Beocenter 2500 156, 157, 163, 169
Beocenter 2600 76, 99
Beocenter 2800 76, 77
Beocenter 3300 76
Beocenter 3500 74, 76, 78, 181, 182
Beocenter 3600 76, 77
Beocenter 4000
   1970s model 77, 78, 143, 206
   1980s model 101
Beocenter 4600 76, 77, 99
Beocenter 5000 110
Beocenter 7000 110, 111, 114, 213
Beocenter 7002 110, 111, 213
Beocenter 7007 111, 211, 213
Beocenter 7700 110, 111, 112, 116, 117, 119, 211, 213
Beocenter 8000 152
Beocenter 8500 151, 152, 169
Beocenter 9000 78, 137, 138, 141, 143, 144, 147
BeoCenter 9300 169, 170, 177
BeoCenter 9500 147, 148, 150, 151, 163, 169
BeoCenter AV5 174, 175, 176
Beocenter Terminal 112
Beocom 2000 145, 146, 170, 172
Beocom 2400 171

Beocom 5000 172
BeoCom 6000 170, 171, 172
Beocom 9000 170, 172
Beocom 9500 172
Beocom mobile telephones 170
Beocord 84U 14, 15
Beocord 900 71, 74
Beocord 1100
   cassette model 58, 71, 72, 74, 75
   open-reel model 27, 29
Beocord 1101 71
Beocord 1200 40, 70, 71
Beocord 1500
   open-reel model 29, 41
Beocord 1600
   cassette model 104
   open-reel model 70, 71
Beocord 1700
   1970s model 53, 71, 116
   1980s model 103
Beocord 1800 41, 71
Beocord 1900 58, 59, 103, 209
Beocord 2000
   De Luxe, open-reel model 27, 28, 29, 41, 70, 73
   cassette model 100, 105
Beocord 2200 7, 71, 72
Beocord 2400
   cassette model 103
   open-reel model 41, 43, 70, 71
Beocord 3500 152
Beocord 4000 video recorder 97, 218
Beocord 4500 138, 152
Beocord 5000
   1970s model 67, 72, 73, 77, 206
   1980s model 118, 190
Beocord 5500 144
Beocord 6000 114
Beocord 6002 114
Beocord 8000 92, 93, 106, 109, 113, 114
Beocord 8002 106, 107, 109, 114
Beocord 8004 92, 108, 109, 114, 115, 118
Beocord 8800V 98, 200
Beocord 8802V 98
Beocord 9000 107, 108, 109, 115, 115, 125
Beocord Belcanto 20, 21
Beocord Stereomaster 27
Beocord V 3000 154, 159, 160
Beocord V 6000 160, 167, 177
Beocord V 8000 177
Beocord VCR 60 98
Beocord VHS 63 126
Beocord VHS 66 126, 189, 210

Beocord VHS 80 124, 125, 126
Beocord VHS 82 126, 131
Beocord VHS 82.2 154
Beocord VHS 90 124, 125, 154
Beocord VHS 91 125, 126, 210
Beocord VHS 91.2 154, 210
Beocord VX 5000 126, 127, 129, 130, 132, 145, 153, 154, 219
Beocord VX 5500 153, 154, 219
Beocord VX 7000 159, 160, 161, 168, 177, 219
Beogram 1000 29, 31, 53, 63, 68, 192, 209
Beogram 1001 53, 68
Beogram 1100 68, 69, 76, 76
Beogram 1102 70
Beogram 1200 40, 41, 42, 181, 182, 205
Beogram 1202 68, 74
Beogram 1500 59, 70
Beogram 1700 69, 70, 191
Beogram 1800
   1960s/70s model 40
   1980s model 100, 106, 118
Beogram 1900 57, 69, 209
Beogram 1902 70, 76
Beogram 2000
   1970s model 68, 199
   1980s model 106
Beogram 2200 70
Beogram 2400 59, 70, 114
Beogram 3000
   1960s model, Acoustical 37
   1960s model, Thorens 37
   1970s model 68, 74
   1980s model 106, 144
Beogram 3300 106, 144
Beogram 3400 57, 69
Beogram 4000 7, 37, 62, 64, 65, 66, 67, 68, 109, 119, 181, 192
Beogram 4002 66, 67, 68, 70
Beogram 4004 59, 68
Beogram 4500 157
Beogram 5000 118, 119
Beogram 5005 119, 138
Beogram 5500 138
Beogram 6000
   1970s CD4 model 67, 68
   1980s radial model 114
Beogram 6002 114
Beogram 6006 114, 192
Beogram 7000 119
Beogram 8000 93, 94, 109, 114, 119, 192
Beogram 8002 109, 111, 114, 115, 192
Beogram 9000 144
Beogram 9500 149

INDEX

Beogram CD 50 119, 133, 134, 135, 136, 137, 138, 191, 211, 213
Beogram CD 3300 136, 136, 138
Beogram CD 4500 138
Beogram CD 5500 137, 138
Beogram CD X 135, 136, 137, 138, 191
Beogram CDX 2 136
Beogram RX 2 118
Beogram TX 115
Beolab 1 177
Beolab 150 141
Beolab 200 147
Beolab 1700
 Amplifier 53, 54
 System 52, 53, 71, 116
Beolab 2500 157, 158, 159, 162, 166, 170, 177
Beolab 3000 147, 157, 163, 170
BeoLab 4000 176, 177
Beolab 4500 170, 177
Beolab 5000
 amplifier 34, 36, 37, 42
 loudspeaker 147, 157, 163
 system 34, 36, 37, 44, 53, 63, 181, 209
Beolab 6000
 loudspeaker 155, 162, 163, 164, 170, 194, 200
 system 113, 114
Beolab 8000
 loudspeaker 162, 163, 170, 174, 194
 system 90, 94, 104, 106, 109, 110, 113, 114, 120, 211
Beolab Penta 140, 141, 142, 147, 157, 162, 170, 177, 194, 206
Beolab Terminal 90
Beolink 1000 56, 132, 139, 140, 144, 147, 149, 151, 154, 159, 166, 167, 172, 219
Beolink 5000 149, 150, 151, 161, 167, 168, 219
Beolink 7000 149, 150, 151, 219
Beolit 39 14
Beolit 41 13
Beolit 400 45, 46, 47
Beolit 500
 Colouradio model 45, 47, 48
 original model 31, 32, 33, 114, 187
Beolit 600
 Colouradio model 45, 47
 original model 20, 181
Beolit 700
 Colouradio model 45, 46, 47, 48
 original model 22, 23
Beolit 707 101, 102
Beolit 800 23, 24
Beolit 1000 33, 34
Beolit Teena 19
Beomaster 700 26
Beomaster 800 51
Beomaster 900 24, 25, 26, 27, 31, 42, 62, 182
Beomaster 901 6, 49, 50, 51, 53, 74
Beomaster 1000 29, 30, 31, 209
Beomaster 1001 53, 209
Beomaster 1100 49, 53, 74, 75, 76
Beomaster 1200 40, 41, 53
Beomaster 1400 26, 42, 62
Beomaster 1500 61, 62, 76, 103

Beomaster 1600
 1980s model 104
 original model 26, 42
Beomaster 1700
 1970s model 53
 1980s model 103
Beomaster 1900 57, 58, 60, 69, 70, 104, 105, 138, 147, 165, 182, 200, 209
Beomaster 2000
 1970s model 54, 55, 57
 1980s model 104, 105, 118
Beomaster 2200 60, 61, 76, 208
Beomaster 2300 105
Beomaster 2400 57, 59, 60, 70, 104, 105
Beomaster 2400-2 59, 103
Beomaster 3000-2 49, 50, 74
Beomaster 3000
 1960s/1970s model 42, 44, 49, 63, 181
 1980s model 104, 105, 211
Beomaster 3300 144
Beomaster 3400 57
Beomaster 4000 49, 53
Beomaster 4400 49, 50, 53, 61, 62, 113, 114, 189, 195
Beomaster 4401 53, 62
Beomaster 5000
 1960s model 34, 35, 36, 38, 42, 44
 1980s model 116, 117, 134
Beomaster 5500 140, 141, 144, 219
Beomaster 6000 commander 55, 56
Beomaster 6000
 1980s model 113, 114, 115, 140, 141, 211
 4channel 53, 54, 55, 56, 67, 72, 87, 103
Beomaster 6500 219
Beomaster 7000 219
Beomaster 8000 90, 91, 92, 103, 109, 113, 114, 115, 120, 140, 141, 156, 211
Beomaster Control Module 59
Beomic BM 3 15
Beomic BM 5 15, 28, 187
BeoSound 9000 172, 173, 174, 181, 210
BeoSound Century 165, 166, 169, 210
BeoSound Ouverture 169, 170
Beosystem 10 101, 102, 103, 166
Beosystem 1200 40, 41
Beosystem 1500 213
Beosystem 2000 105, 136, 137
Beosystem 2300 158
Beosystem 2500 154, 155, 156, 157, 158, 161
Beosystem 3000 105, 136, 137
Beosystem 3300 106, 137, 147
Beosystem 3500 151, 154
Beosystem 4500 138, 147, 151, 152, 157, 165
Beosystem 5000 115, 116, 117, 118, 119, 120, 126, 134, 144, 145, 155, 169, 182
Beosystem 5500 138, 139, 140, 145, 147
Beosystem 6000 115
Beosystem 6500 147, 149

Beosystem 7000 150, 169
Beosystem 9500 149
Beosytsem 5000 133
BeoVision 1 177, 178
Beovision 600 84, 85
Beovision 601 84, 85
Beovision 1400 39
Beovision 1600 85
Beovision 2600 82, 83, 182
Beovision 3000, 3300, 3800 & 3900 95
Beovision 3000
 original model 38, 39, 82, 83, 84, 159, 182
Beovision 3100 83
Beovision 3200 82, 83
Beovision 33XX series 96, 129
Beovision 3400 83, 84, 85, 86
Beovision 3500 85, 86, 87
Beovision 3502 87, 88
Beovision 3802 87, 88, 182
Beovision 4402 87, 88, 183
Beovision 5500 96
Beovision 5502 129, 131
Beovision 6000 55, 85, 86, 88
Beovision 6000 Commander 87
Beovision 6002 88, 97
Beovision 7700 96
Beovision 77XX series 129
Beovision 7800 97, 128
Beovision 7802 125, 128, 129, 159
Beovision 8800 95, 96, 98, 188
Beovision 8900 97, 128
Beovision 8902 125, 129
Beovision 9000 97
BeoVision Avant 166, 167, 168, 169, 175, 176, 177, 201, 202
Beovision L 2500 129
Beovision L 2800 129
Beovision LE / ME series 159
Beovision LE 6000 159
Beovision LS 6000 159
Beovision LS series 154
Beovision LX 2500 129
Beovision LX 2502 127, 153
Beovision LX 2800 129
Beovision LX 2802 127, 129, 130, 153, 182
Beovision LX 4500 153
Beovision LX 5000 159
Beovision LX 5500 153
Beovision LX 6000 159, 168
Beovision M 20 131
Beovision ME 6000 159
Beovision MS 6000 159
Beovision MX 1500 132, 153
Beovision MX 2000 126, 131, 132
Beovision MX 3000 132, 153, 218
Beovision MX 3500 153, 210
Beovision MX 4000 158, 159
Beovision MX 4500 132, 219
Beovision MX 5000 127, 132, 153, 219
Beovision MX 5500 153, 210
Beovision MX 6000 159, 161
Beovision MX 7000 158, 159, 168
Beovision Video Terminal 96, 98, 129, 218
Beovox 1001 53, 209
Beovox 1200 40
Beovox 1600 31, 193

INDEX

Beovox 2500 cube 36, 37
Beovox 2702 193
Beovox 3000
    1990s panel model 147
Beovox 5000
    1960s model 36, 37, 209
    1990s panel model 147
Beovox 5700 79
Beovox C 30 80, 81
Beovox C 40 80, 81
Beovox C 75 80
Beovox C series miniature loudspeakers
    80, 81, 82
Beovox Cona 146
Beovox CX 50 80, 81, 177, 193, 206
Beovox CX 100 80, 81, 112, 158, 177
Beovox CX series miniature loudspeakers
    80, 81, 82, 102, 146, 194
Beovox M & MS series Uniphase
    loudspeakers 80, 120
Beovox M 70 80
Beovox M 100 80
Beovox M 150 120
Beovox MC 120.2 121
Beovox P 30 80
Beovox P 45 80
Beovox P series Uniphase loudspeakers
    80
Beovox Penta 141, 142
Beovox RL 35 124
Beovox RL 45 122, 124, 194
Beovox RL 45.2 124
Beovox RL 60 122, 124, 194, 206
Beovox RL 60.2 122, 124, 206
Beovox RL 140 122, 124, 215
Beovox RL 6000 177
Beovox RL series loudspeakers See Red
    Line loudspeakers
Beovox S 22 76, 80, 206
Beovox S 25 80
Beovox S 30 80, 99
Beovox S 35 194
Beovox S 35.2 194
Beovox S 45
    1970s model 80
    1980s model 121
Beovox S 45.2 80
Beovox S 55 120, 121, 215
Beovox S 60 80
Beovox S 70 80
Beovox S 80 121, 206
Beovox S 80.2 79, 121
Beovox S 120 121
Beovox S series Uniphase loudspeakers
    80
Beovox X 25 100
Beovox X 35 105
Beta video format 97, 124
Bofa 11
Bottle opener 12
BSR 20, 21
cassette deck adjustment 205, 206
Capri models 16
CCC system 107, 108, 109
CD Terminal 135, 211
CD-i video format 175, 176
CD-R & CD-RW discs 191
CD4 See quadraphonic
Colouradio range 44, 47, 181, 182
Compact Cassette audio format 70

CP90 TV chassis 132
Dance-proof turntables 65
Danish TV boom 16
DAT audio format 174
Datalink 92, 104, 106, 114, 117, 118,
    125, 137, 209, 210, 211, 213
DCC audio format 174
DECT telephone system 172
Denon 90
Deutsche Grammophon 191
digital TV set-top boxes 219
Dirigent 17
Dolby B noise reduction 71, 73
Dolby C noise reduction 108, 109, 144
Dolby HX Pro 101, 106, 107
Dolby Pro Logic 161, 168
dual standard TV sets 39
Duus Hansen, Lorenz 14, 15
DVD video format 133, 175, 219
EKCO 11
electronic covers 144
FA 1 antenna 207
Feel Commander 59, 88, 89
foam roll edges 82, 121, 124, 143, 193
foam rot 121, 143, 194
Form 1 headphones 123, 124, 177
Garrard 26
Grand Prix models 16
Grundig 160
headphones 123, 124
Hitachi 101, 109, 124, 125, 126, 131,
    143, 154
HX Pro See Dolby HX Pro
hybrid chip amplifier 99, 104, 147
Hyperbo 12
ICC series TV chassis 131, 153
ICEpower 177
in line picture tube 87
in-house cassette mechanism 72, 77
indoor radio aerials 208
integrated pickups 63
ITT 131
JVC 57, 67, 97, 124
Jensen, Jacob 31, 35, 49, 63, 74, 78, 96
Laser Vision video disc format 133
Lewis, David 96, 131
light controls 162
Linn Sondek 67
Master Control Panel 5000 117, 119,
    134
Master Control Panel 5500 139, 148
Master Control Panel 6500 148, 149
Master Control Panel 7700 112, 117
Master Panel AV 9000 161, 169
Matchline 160
Mini 610 De Luxe 17
MiniDisc audio format 174, 211
MMC 1 > 5 pickups 109, 111, 119
MMC 20 series pickups 69, 109
MMC pickups 63, 64, 68, 69, 76, 191
Moldenhawer, Henning 25, 26, 88, 96
multiroom system 111, 112, 119, 139,
    144, 177
Museum of Modern Art 40
N1500 video format 97
NICAM 129, 153
Nordmende 131
Olufsen, Svend 9, 10, 12
open-reel tape recorders 41, 70, 71, 97,
    206, 218

PCM digital audio format 133
Permanent Colour Truth 86
Phase 2 Delta colour tube 83
Philips 22, 39, 66, 70, 83, 85, 87, 95,
    97, 98, 101, 102, 126, 129, 132,
    133, 135, 136, 137, 138, 144,
    153, 157, 160, 167, 169, 174, 175,
    191
PIN code 186
Power Link 146, 147, 152, 153, 157,
    161, 162, 173, 213, 214
Pramanik, Subir 63
Profeel 160
quadraphonic 54, 55, 57, 67, 68, 69,
    161
Radio Heart 12
Rank 22
RCA 42, 44, 87
rarity 182
razors 14
Red Line loudspeakers 122, 123, 124,
    194, 215
Sanyo 147
Second Life scheme 7, 183
serial numbers 185, 186
Sharp 102
shavers See razors
Sony 39, 87, 90, 97, 101, 102, 107,
    124, 133, 160, 166, 177
SP series magnetic pickup 16, 42, 63,
    191, 192
Speaker Link 214
SQ format See quadraphonic
stack systems 53, 115, 116, 118, 120,
    143
tangential drive 94, 114, 119, 192
tangential tracking 63, 94, 119, 192
TDA1541 DAC chip 137, 157
TDK MA-R cassette 108
teak oil 196, 197, 198
Technics 90
Telephone Book 14
Teletext 95, 96, 129, 154, 160, 167
Terminal 3300 139, 140
Terminal 5000 117
Thorens 37
Toshiba 87, 131
transit locks 186
turntable adjustment 204, 205
two-way remote control 111, 147, 150,
    152, 153, 167, 168, 219
U 70 headphones 123, 124
ultrasonic remote controls 55, 59, 86,
    87, 88, 89
Uniphase loudspeakers 79, 120, 121,
    122, 194, 215
universal recording machine 124,
    125
V Terminal 138
V2000 video format 97, 98, 124
valves 11, 12, 20, 21, 24, 18, 39, 83
VHS video format 124
Video Terminal See Beovision Video
    Terminal
VX sensor 127
Whiteline 182
widescreen television 166
wire recorders 15
wood care 195
Zeuthen, Gustav 63